The RING of GYGES

What We Become When No One Is Watching

Ryan Gross

ISBN 978-1-970970-07-4

Disclaimer

This book is a work of narrative non-fiction. All events, individuals, organizations, legal proceedings, and research findings described herein are based on publicly available sources, including but not limited to court records, government documents, congressional testimony, peer reviewed academic research, investigative journalism, declassified intelligence materials, and published interviews.

No classified or privileged information was used in the preparation of this book. All sources are documented in the References section.

The author has made every effort to ensure the accuracy of the information presented. However, this book is intended as a work of analysis and synthesis, not as legal, political, or policy advice. The interpretations and conclusions drawn from the evidence are the author's own and do not represent the views of any institution, organization, or government agency. Some of the events described in this book involve ongoing legal proceedings, evolving policy frameworks, and contested historical narratives. Readers are encouraged to consult the original sources for the most current information.

The inclusion of any individual, organization, or government program in this book does not constitute an accusation of wrongdoing beyond what has been established through the cited public record.

DEDICATION

For everyone who stood in the dark and chose to do the right thing.
Integrity is everything.

Author Note

In 375 BCE, the Greek philosopher Plato told a story about a shepherd who found a ring that made him invisible. The shepherd used the ring to seduce a queen, murder a king, and seize a throne. Plato's brother Glaucon used the story to argue that no human being is willingly moral. That justice is nothing more than fear dressed in respectable clothing. That given the power to act without consequence, every person alive would become a predator. Socrates disagreed. To this day, the debate has yet to be settled.

The internet gave billions of people the ability to act anonymously. Corporations built structures that shielded individual decision makers from the consequences of their choices. Governments classified programs that would never survive public scrutiny. Financial systems made it possible to move vast fortunes without a name attached. And in every case, the pattern held. When people believed they were invisible, their behavior changed. Sometimes for the worse. Sometimes, remarkably, for the better. The question that drove this book is not whether Glaucon or Socrates were right. The evidence suggests they were both partially correct, which is far more interesting than either of them being fully vindicated. The real question is what we do with that knowledge. We live in a world where the ring cannot be un-invented. Neither can the camera. The tools of invisibility and the tools of surveillance will only grow more powerful. The question is whether we can build societies, institutions, and people who do not require constant observation to act with decency.

I do not pretend to have a definitive answer. What I have is evidence. This book presents that evidence as honestly as I can, drawn from history, psychology, investigative journalism, court records, leaked documents, and peer reviewed research. Where sources conflict, I have noted the disagreement. Where the evidence is incomplete, I have said so. Where my own perspective shapes the interpretation, I trust the reader to weigh it accordingly.

Every claim of fact in this book is supported by sources documented in the References section. Inline citation numbers throughout the text correspond to specific references organized by chapter. I encourage skeptical readers to check my work. A book about accountability should be accountable. Yet, you will see my opinions in a few of them. That's what they are... opinions.

One final note. This is <u>not</u> an anti-technology book. It is <u>not</u> an anti-government book. It is <u>not</u> an anti-corporate book. It is a book about human nature and the structures we build around it. Technology, government, and corporate power are neither good nor evil in themselves.

How Best to Read This book

The book is laid out in a way to help slowly immerse the reader into the story of Gyges in Book II of Plato's Republic. And how a common theme can be seen in today's world. The book will provide differing viewpoints from ancient history, traveling into modern voices that you may already have heard of. To do this the first few chapters provide those viewpoints and in such add context. If you are already familiar with the story, feel free to just jump right to chapter three. Each chapter is meant to highlight a particular aspect that feeds the narrative. The Ring of Gyges is still alive today as it ever was.

Prologue - The Shepherd's Cave

The story of Gyges and his ring appears in Book II of Plato's *Republic*, written around 375 BCE. The tale ranks among the most famous thought experiments in Western philosophy. Crucially, it is not an argument that Plato himself endorses.[1] This is how it unfolds...

The hillside was quiet that morning. Grass bent under a warm Lydian wind, with a shepherd who sat among his flock on a slope overlooking the Hermus valley. He was an unremarkable man. His name was Gyges, and he worked for the king of Lydia, tending sheep on land that belonged to the crown. The work was dull and repetitive. Gyges knew the rhythm of the seasons by the thickness of wool under his hands. He knew the stars by their positions over the ridge at nightfall, and very little else beyond that.

The storm came without warning, darkening the sky over the Hermus valley. Then the earthquake followed. The ground shuddered once, then lurched with a violence that sent the flock scattering across the hillside. Gyges fell to his knees. A sound rose from beneath the earth, something deep and grinding, as though the world itself were splitting along a hidden seam. The soil cracked open. A jagged chasm appeared in the hillside where solid ground had been only moments before. Dust rose in thick clouds from the opening. Loose stones tumbled inward and vanished into darkness, their impacts echoing for far too long before fading into silence.

The tremor subsided. The sheep regrouped at a cautious distance. Gyges stood at the edge of the fissure and stared into it. The opening was wide enough for a man to pass through. A crude stairway of broken rock descended into the earth at a steep angle. Light from the surface reached perhaps twenty feet before the passage bent and swallowed everything in shadow.[1]

He ventured down. The air changed within the first few steps. The warmth of the Lydian afternoon gave way to a cool dampness that smelled of stone and age. The walls were not natural rock. They had been shaped by hands, carved in smooth planes that caught the faint light filtering from above. Gyges moved deeper. The passage opened into a chamber, and in the center of that chamber stood something that stopped him where he stood.

It was a horse. A hollow bronze horse, large enough to hold a man inside its body. The metal surface had gone green with centuries of oxidation. The legs were thick as columns, and the eye sockets stared out hollow and dark. The craftsmanship was extraordinary, detailed down to the veins along the neck and the individual strands of the mane. No shepherd in Lydia had ever seen anything like it. No one alive likely had either.

Gyges circled the bronze horse. He found a seam in its side, a panel that swung open at his touch. He crouched and looked inside. A corpse lay within the hollow body. The figure was enormous, larger than any human Gyges had ever seen, with limbs that seemed to belong to a different order of being entirely. The flesh had long since dried and drawn tight against the bone. There was no clothing on the body. It carried no weapons and bore no insignia. On one of its massive fingers sat a single gold ring.

The ring was beautiful. The band was thick and polished, with a bezel carved in a design Gyges could not identify. The gold caught what little light reached the chamber and seemed to hold it, glowing faintly against the darkness of the dead hand. Gyges looked at the corpse, then at the ring. He reached out, worked it free from the stiffened finger, and slipped it onto his own hand. Then he climbed back out of the earth.[1]

Several days later the shepherds of the region gathered, as they did each month, to prepare their reports for the king. Gyges sat among them. The ring was on his finger. He had not removed it since the cave.

During the meeting he fidgeted with the band, turning it absently while the others spoke. At one point he rotated the bezel inward, toward his palm. The conversation continued around him, but something in his character changed. The man beside him shifted his weight to fill a gap. Another shepherd gestured toward the space where Gyges sat and spoke about him as though he were not there. He spoke about Gyges as though Gyges had already gone home.

Gyges sat perfectly still, not breathing loudly, not shifting or speaking. The shepherds around him continued their discussion with the easy openness of men who believed themselves unobserved. They made no effort to include him because they could not see him. He was sitting among them, close enough to touch, but he had become invisible.

Gyges turned the bezel outward. The man beside him startled. Heads turned. Someone asked when he had arrived. Gyges said nothing of consequence. When he rotated the bezel inward again the shepherds resumed their conversation as if a chair had simply emptied. A second rotation outward brought him back into view. The reactions confirmed what he already suspected. The ring was not a trick of light or a lapse in attention. It granted its wearer the power to vanish entirely from the perception of other human beings.[1]

He sat with that knowledge for a long moment. The weight of it settled over him slowly, the way understanding sometimes does. What he felt was not fear, and it was not joy. It was the particular stillness of a man realizing that every boundary he had ever known... every rule, every expectation, every consequence that had shaped his life from the day he was born now ceased to apply. Then he decided what to do with it.

Gyges arranged to be included among the messengers sent from the shepherds to the court of King Candaules. This was not difficult. The assignment rotated regularly, and a man who expressed willingness was rarely turned away from an errand no one else wanted. He traveled to the capital and entered the palace as one anonymous servant among many.

At court he used the ring. Unseen, he moved through private chambers and listened to conversations that were never meant for his ears. Over the course of days he learned the layout of the palace, the schedules of the guards, and the habits of the queen. No other subject in the kingdom could have studied the architecture of power from such a vantage point. Gyges was invisible, and invisibility made him patient.

The queen was his first target. The historical accounts do not specify how the seduction unfolded.[2] What Plato's telling makes clear is that Gyges did seduce her, and that the seduction was strategic rather than romantic.[1] He needed an ally inside the palace walls, someone whose position granted authority and whose grievances could be channeled toward his purposes. The queen provided both. She was close to the king and trusted by the court. In time she became Gyges' co-conspirator without the court ever knowing he existed as anything more than a shepherd who delivered reports.

The conspiracy was efficient. Gyges and the queen identified the moment of greatest vulnerability in the king's routine. Together they planned the method and chose the location. When the night came, Gyges used the ring to enter the king's private chambers without detection. He killed Candaules. The murder was quick. The palace learned of the king's death before anyone thought to look for the man responsible, and by then the queen had already named Gyges as the rightful successor. He took the throne of Lydia.[1],[2]

The progression deserves attention because it reveals something about the structure of corruption itself. In Plato's telling, Gyges does not agonize. The discovery of invisibility leads swiftly and without hesitation to action. He arranges a visit to the palace, seduces the queen, murders the king, and takes the throne. Each step is worse than the one before it, and each step is made possible only by the one that preceded it. Observation enabled seduction, which enabled conspiracy, which enabled assassination, which delivered a kingship that no shepherd had any right to claim. At no point did Gyges face a consequence that might have interrupted the sequence. Nobody saw him, questioned him, or held him accountable. The ring removed every external check on his behavior, and without those checks, an ordinary shepherd became a predator whose path from invisibility to the throne unfolded with a swiftness that is itself part of the parable's warning.

Glaucon, one of Plato's older brothers, tells it during a long conversation about the nature of justice. Glaucon presents the tale as a provocation. His claim is not that Gyges was right, but that Gyges was typical. The argument he makes to Socrates is direct and unsettling. He states it without equivocation.

No one is willingly good. People behave justly only because they lack the power to behave otherwise. Morality is not an internal quality of the soul. It is an external constraint imposed by the fear of being caught and punished. Justice is a compromise that the weak accept because they cannot afford to be unjust. The strong accept it only because they cannot afford to be seen as unjust. Remove the possibility of being seen, and the distinction between a just man and an unjust man disappears entirely. Give both men a ring of Gyges, and they will act the same way. The just man will seduce, steal and kill no differently than the unjust man. Because justice was never about character in the first place. It was always about visibility.[1]

Glaucon pushes the argument further. He asks Socrates to consider two lives. In the first life, a perfectly unjust man receives the reputation of perfect justice. Everyone believes him to be good. He is honored, trusted, and rewarded with every advantage society can offer. In the second life, a perfectly just man receives the reputation of perfect injustice. The world believes him to be corrupt. He is despised, imprisoned, and tortured. Glaucon asks the question that sits at the heart of moral philosophy... which man is better off? If justice is truly valuable for its own sake, then the just man should be happier even while suffering. If the unjust man with the good reputation is happier, then justice is nothing more than a tool for managing appearances.[1]

Socrates takes the challenge seriously. His response occupies the remaining eight books of the *Republic*. He constructs an elaborate analogy between the individual soul and the ideal city-state, arguing that justice is a form of internal harmony. The just soul is ordered properly, with reason governing spirit and appetite in a balanced hierarchy. An unjust soul is disordered regardless of external success. The tyrant, Socrates insists, is the most miserable of all men. Even if he sits on a throne surrounded by wealth and praise, his soul is a chaos of appetites and fears. He is enslaved by desires that can never be satisfied. The ring of Gyges does not liberate its wearer. It traps him in a prison of his own appetites.[1]

The response is brilliant. It is also, after centuries of debate, far from settled. Philosophers have argued about it ever since. Some find Socrates persuasive. Others find his ideal city authoritarian and his psychology outdated. The question Glaucon asked remains as alive as the day he asked it.

Would a good person remain good if guaranteed that no one would ever find out?

This book is not about who was right. Plato and Glaucon debate matters because it is easier to see it being played out. Because we are running the experiment every day now. For most of human history the Ring of Gyges was a thought experiment and nothing more. Invisibility was a fantasy. People lived in small communities where everyone knew their names, their families, and their reputations. The village was its own surveillance apparatus. The cost of misbehavior was exile, shame, or violence. The ancient world had no shortage of cruelty, but the cruelty was rarely anonymous. When a man wronged his neighbor, both parties knew exactly who was involved.

That world is gone. It did not vanish all at once, but it has vanished thoroughly. Over the past several decades the human species has constructed an elaborate network of systems that grant varying degrees of invisibility to billions of people simultaneously. The internet was the first and most obvious ring. It gave ordinary individuals the ability to speak, act, and transact without revealing their identities. A teenager in a bedroom could reach millions of strangers while remaining as faceless as Gyges among the shepherds. Early internet theorists celebrated this anonymity as liberation. It would free speech from censorship, empower dissidents, and protect whistleblowers. It would level hierarchies and democratize information. All of that proved true. Yet all of it also proved insufficient as a description of what anonymity actually does to people.

The internet is only one ring among many. Corporate structures provide their own form of invisibility. The modern corporation diffuses responsibility across so many layers of management, legal entities, and subsidiary relationships that individual decision-makers become effectively invisible. A pharmaceutical executive who prices a lifesaving drug beyond the reach of the sick is shielded by the same structural opacity that protected Gyges in the king's chamber. The mechanism is different. The outcome is remarkably similar.

Government secrecy operates by the same principle but on a far larger scale. Classification systems, intelligence agencies, and national security doctrines create vast zones of activity where public oversight cannot reach. When a government acts in secret, the officials involved are wearing the ring. They are making decisions that affect millions of lives while remaining invisible to the people those decisions affect. The results of those decisions eventually surface, sometimes after years or decades, but the moment of choice happens in the dark.

Financial anonymity is perhaps the purest modern analogue to the ring itself. Shell companies, offshore accounts, cryptocurrency, and opaque banking systems allow wealth to move across the globe without revealing who is moving it or why. Money launders identity as effectively as any bronze-age artifact. The person behind the transaction vanishes, leaving behind nothing but numbers on a screen. Only the transaction remains, stripped of human fingerprints and human accountability.

Billions of people now have access to one or more of these rings. Most do not think of their anonymity in philosophical terms. They do not invoke Plato when they create a burner account or route money through an offshore entity. Yet the dynamic is the same. The experiment Glaucon proposed as a hypothetical is being conducted at civilization scale, with no control group and no informed consent. The results are accumulating faster than any philosopher or social scientist can catalog them.

This book examines those results. It is organized into six parts. Part One, *The Ancient Ring*, revisits the philosophical origins of the question, from the original story of Gyges to the debate it has provoked among thinkers from Plato to the present. Part Two, *The Digital Ring*, explores what happens when individuals gain anonymity online, from the earliest internet forums to the algorithmic platforms that now shape public discourse. Part Three, *The Institutional Ring*, follows the ring into the world of corporations and finance, examining how structural invisibility and financial anonymity enable corruption on a scale that would have staggered even the most cynical ancient philosopher. Part Four, *The State's Ring*, turns to governments and looks at what happens when state power operates behind walls of secrecy, classification, and surveillance. Part Five, *The Watchers*, considers the opposite problem. What happens when the ring is taken away entirely, when surveillance becomes total and privacy disappears. The sixth and final part, *The Evidence and the Choice*, asks whether the experiment has produced any evidence of genuine moral behavior, actions taken in the dark by people who chose to be good even when no one was watching, and offers a framework for thinking about the choices ahead.

Each chapter tells a story. Some of the stories involve famous figures whose names will be immediately recognizable. Others involve ordinary people whose decisions, made in moments of total invisibility, reveal something essential about human nature. The chapters draw on philosophy, psychology, history, economics, and investigative journalism. They are grounded in evidence and sourced to the available research. They are also, in the end, about a single question that has haunted humanity for as long as anyone has bothered to write it down.

Gyges climbed out of the cave with a gold ring on his finger and the knowledge that no one could see him. Within months he sat on a throne that was never meant to be his. He arrived there not through talent or birthright but through the simple removal of consequence. The ancient story presents this as a parable about human nature, a cautionary illustration of what we become when the constraints are removed. The modern world has removed the constraints for billions of people at once. A chasm has opened beneath the foundations of civilized life, and the ring has been found. The question is no longer hypothetical. It is empirical. What do the results show?

References

[1] Plato. *Republic*. Translated by G.M.A. Grube, revised by C.D.C. Reeve. Book II, 359a-360d.

[2] Herodotus. *The Histories*. Book I.

Chapter One - The King, the Queen, and the Bodyguard

The story transformed over time from its origin. This chapter will help you see the differences between Herodotus, who wrote his *Histories* in the fifth century BCE and Plato's version in Book II of the *Republic*. It's a historical comparison. This next chapter sets up the argument from ancient times to the recent past including different viewpoints. It really helps set up the foundation of future arguments. If it gets dry you can skip them, but it helps cement the context of the book.

The Pactolus River carried gold in its current. The gold was not metaphorical or poetic. Actual flecks of electrum tumbled along the riverbed, glinting in the Anatolian sun as the water wound its way past the city walls.[1] The residents of Sardis had been pulling wealth from that river for generations before anyone thought to stamp it into coins. They waded into the shallows with sheepskins spread across the gravel, trapping the tiny bright particles in the wool, wringing out fortunes with their hands. In the seventh century before Christ, this was one of the richest places on earth, and almost no one in the Western world has heard of it.

Sardis sat in what is now western Turkey, at the edge of a broad fertile plain where the Hermus River valley opens toward the Aegean Sea. The Tmolus mountain range rose sharply to the south, its peaks still carrying snow well into spring. The citadel of the city perched on a spur of nearly vertical rock that made the acropolis all but impossible to assault from three sides.[3] The lower city spread out below in terraced layers, its streets lined with workshops and merchant stalls, its marketplaces thick with the sound of a dozen languages spoken at once. Caravans arrived from the east carrying lapis lazuli, textiles, and spices along routes that would later become the Persian Royal Road. Greek traders pushed inland from their coastal settlements to the west, bringing olive oil, painted pottery, and an insatiable appetite for Lydian gold. Sardis was the hinge between two civilizations. Everything that moved between the Greek world and the empires of Mesopotamia passed through or near this city, and the Lydians collected a share of every transaction.[5]

The Harvard-Cornell Archaeological Expedition, which began excavating the site in 1958 and continued for decades afterward, revealed just how sophisticated this place had been. The team uncovered massive fortification walls built from limestone blocks that still fit together with remarkable precision after twenty-seven centuries of earthquakes and erosion. They found gold-refining installations along the Pactolus, including furnaces and cementation vessels designed to separate gold from silver at the molecular level.[3] They documented evidence of a thriving commercial district that predated the classical Greek city-states by a century or more, with workshops for metalwork, textile production, and ceramics all clustered in a dense urban core. The sheer scale of the excavation told its own story. Sardis was not a village that had gotten lucky with geography. It was a planned urban center with infrastructure that rivaled anything else in the seventh-century Mediterranean.[5] The Lydians were not a footnote in someone else's history. They were innovators operating at the frontier of ancient technology. They may have been the first people on earth to mint coins from metal, an invention so consequential that it reshaped the entire concept of value and ultimately made possible the global financial systems we use today.[4] The gold-bearing sands of the Pactolus were the raw material for that revolution.

Around 680 BCE, a man named Candaules sat on the Lydian throne. He was the last king of the Heraclid dynasty, a lineage that claimed descent from the demigod Heracles himself and had supposedly ruled Lydia for twenty-two generations.[1] Whether that genealogy was real or invented is beside the point. What mattered was that Candaules wielded absolute power in a kingdom that controlled one of the most strategic crossroads in the ancient world. He commanded armies that could march west to the Aegean or east toward the plateau of central Anatolia. He collected tribute from the cities and villages that depended on Lydian trade routes for their survival. He entertained ambassadors from Greece and from the great kingdoms to the east, hosting them in a palace that sat atop the fortified acropolis overlooking the fertile valley below. The Heraclid kings had ruled this kingdom for so long that their legitimacy felt as permanent as the mountain range on the southern horizon. By every measure available to the ancient world, Candaules was a man who had everything a man could possess.

Among the men in his inner circle was a soldier named Gyges. The historical sources are sparse on his origins, but what survives tells us he was a member of the king's personal bodyguard.[1] This was not a ceremonial post. The bodyguard of a Lydian king lived in intimate proximity to the most powerful man in the kingdom. He stood in the corridors of the palace at hours when other men slept. He knew the rhythms of the royal household, which doors opened when, who passed through them, and what happened behind them once they closed. He ate meals within earshot of the king's private conversations. He carried weapons in rooms where other men were forbidden to be armed. He was trusted with the life of the king, which meant he was trusted with the thing the king valued most. That trust was about to be tested in a way that no one could have predicted.

The ancient Greek historian Herodotus, writing roughly two centuries after these events, preserved the story in the opening book of his *Histories*. Herodotus was not a dispassionate chronicler cataloging dates and treaties. He was a storyteller who understood that history lives in the choices people make when circumstances become impossible. He traveled the Mediterranean world collecting accounts from local informants, temple priests, and descendants of the people whose stories he was recording. The narrative he assembled reads less like a historical record and more like the first act of a tragedy. He opened his account of the Lydian kingdom with this particular episode for a reason, placing it before the wars and the treaties and the catalogues of kings.[1] He wanted his readers to understand that the rise and fall of empires often begins not with geopolitics or military strategy but with the private failures of the people in charge. History is shaped in bedrooms as often as it is shaped on battlefields.

Candaules had an obsession. According to Herodotus, the king was consumed by the belief that his wife was the most beautiful woman alive.[1] This was not the quiet private pride of a contented husband. It was a compulsion that demanded external validation. Candaules talked about his wife's beauty constantly. He praised her to his courtiers with a fervor that went beyond the normal boasting of powerful men. He described her to his companions with the insistence of a man who needed the world to confirm what he already believed. The admiration of others was not a bonus for Candaules. It was a requirement. He needed a witness. He chose Gyges.

The king pulled his bodyguard aside and told him that words alone were insufficient. Gyges needed to see the queen with his own eyes. Specifically, he needed to see her unclothed. Candaules proposed a plan with the casual confidence of a man who had never been told no. Gyges would hide behind the door of the royal bedchamber. When the queen entered and undressed for the night, he would watch. Then he would slip out before she noticed, and the matter would be settled.[1]

Gyges recoiled. Herodotus recorded his protest with the kind of direct quotation that suggests this exchange had been passed down through oral tradition for generations before anyone wrote it on papyrus. "Master, what an unsound suggestion," Gyges said. "Are you telling me to look at the queen when she has no clothes on? A woman's honor is lost with her clothing."[1] The protest was not merely polite reluctance from a subordinate playing his expected role. In the cultural context of the ancient Near East, seeing a queen naked was an act loaded with political and religious danger. It blurred the boundary between the king's authority and the king's person. It was a violation of the social order itself, a transgression that could not be undone or walked back once committed. Gyges understood this. He may have understood it better than the king did.

But Candaules insisted. The king dismissed every objection with a wave of his hand. He assured Gyges that the plan was foolproof, that the queen would never know, that there was nothing to fear. He spoke with the absolute certainty of a man who had grown so accustomed to having his desires met that he could no longer conceive of consequences. He could not imagine that a plan of his might fail. He could not imagine that his wife might be more observant than he gave her credit for. The ring of power around his finger was not made of enchanted metal. It was made of unchecked authority, the total and unquestioned certainty that his will could bend reality to his satisfaction.

Gyges had no real choice. When a king gives an order, refusal carries its own set of consequences, and those consequences are rarely survivable. The bodyguard agreed to the plan.

That evening, Candaules led Gyges to the royal bedchamber. He positioned the soldier behind the open door and told him to wait in the shadow between the door and the wall. When the queen entered, she crossed the room and placed her garments on a chair near the door as she undressed for the evening. Her back was turned to the place where Gyges stood watching, his breathing shallow, his body pressed against the stone.[1] He saw what the king wanted him to see. Then he turned to leave, moving as quietly as a man in his position had been trained to move. A bodyguard's footsteps are supposed to be silent. On this night, they were not silent enough. The queen caught the movement from the corner of her eye. She saw Gyges slip through the doorway.[1]

She said nothing that night. She did not cry out in surprise or call for guards. She did not confront her husband in the darkness of the bedroom they shared. She did not weep. She simply noted what had happened. She recognized Gyges, because she had seen him standing in the corridors of the palace a thousand times before. She understood immediately who had arranged this, because she understood the man she was married to. Then she went to bed beside the king who had engineered her humiliation. She lay still in the dark while he slept, and waited.

The restraint is what makes her remarkable. In a story populated by men who act on impulse and appetite, the queen alone demonstrates the lethal patience of someone who understands that the most important response is rarely the first one. She did not react. She calculated. Herodotus did not even record her name, an omission that has frustrated scholars for centuries. Later sources sometimes called her Nyssia, though that attribution is uncertain and may be a later invention.[6] What survives is not her name but her actions, and those actions were devastating in their precision.

The following morning, the queen summoned Gyges to a private audience.[1] She did not send a servant with a vague message or a casual request. She arranged the meeting with the same careful deliberation she had shown the night before. Every detail of the encounter suggests a woman who had spent the dark hours not sleeping but planning, constructing a trap as elegant and as inescapable as the one her husband had built around her. When Gyges arrived, he assumed it was a routine matter. Perhaps the queen needed an escort. Perhaps she had a message to relay. He had no idea that she knew what had happened behind her bedroom door. His confidence in the king's plan had survived the night intact, because men who believe themselves invisible rarely imagine that they have been seen.

The queen ended that confidence with her first sentence. She told

Gyges that she knew what had happened. She told him that her husband had arranged it. Then she gave him a choice, and the architecture of that choice has resonated through twenty-seven centuries of moral philosophy.

She offered two options, and only two. Gyges could die for what he had seen. The queen would have him executed for the violation of her honor, and the matter would be settled with his blood. Alternatively, Gyges could kill Candaules. He could murder the king in his bed, take the throne of Lydia, and marry the queen herself. He would replace the man who had wronged them both.[1]

There was no third option. The queen did not offer negotiation, mercy or a middle path where everyone could pretend the incident had never occurred. She did not suggest that they could all simply agree to forget. She understood something that Candaules had failed to grasp. Certain acts cannot be undone. Once seen, a thing cannot be unseen. The violation had occurred. The only question remaining was who would pay the price for the seeing.

Gyges begged for a third way. Herodotus described him pleading, asking the queen not to force this impossible decision upon him.[1] The soldier who had protested the original plan now protested its consequences with equal desperation. He had not wanted to see the queen naked. He had not wanted any part of this arrangement. He wanted to go back to being a bodyguard, standing in corridors, carrying out his duties in the comfortable anonymity of service. But the queen was immovable. She had been violated, and someone was going to answer for it. The only variable was who.

Consider the geometry of this moment. Candaules, the king, had all factions. Blood had been spilled in the royal bedchamber, and the stain spread outward through the political order like dye in water. Gyges and his opponents ultimately agreed to let the Oracle at Delphi settle the matter. Messengers traveled westward across the Aegean, carrying the question to the most authoritative religious institution in the Greek world. The Oracle confirmed Gyges as the rightful ruler, though the priestess added a prophecy that the Heraclid dynasty would be avenged in the fifth generation.[1] That prophecy would come true roughly a century and a half later, when Croesus, the last of the Mermnad line, fell to Cyrus the Great of Persia.

With divine sanction secured, Gyges ruled Lydia for approximately thirty-six years, from around 680 to 644 BCE.[1] His reign was long and consequential by any standard. He founded the Mermnad dynasty, which would produce some of the most powerful and recognizable kings in the pre-classical world, culminating in Croesus, whose name became a Greek proverb for wealth itself. Gyges expanded Lydian territory through military campaigns against the Greek cities of the Ionian coast, pushing Lydian influence all the way to the shores of the Aegean.[5] He made contact with the Assyrian empire to the east, appearing in Assyrian records under the name Gugu, a detail that confirms his historical existence independently of the Greek literary tradition.[5] He sent ambassadors and tribute to the great powers of his era, forging diplomatic relationships that extended the reach of his kingdom far beyond the Hermus valley.

The tribute he sent to Delphi became legendary in its own right. Herodotus described offerings of gold and silver that were so lavish they were still on display at the sanctuary centuries after Gyges' death, available for any visitor to see and admire.[1] Six golden mixing bowls weighing thirty talents sat in the Corinthian treasury at Delphi, gifts from the former bodyguard to the god Apollo. The Greeks remembered this generosity for generations. Travelers who visited the sanctuary wrote about the Lydian offerings with a mixture of awe and curiosity, wondering at the distant kingdom that could afford to give away so much gold.

The scale of these offerings tells us two things about the man who sent them. First, Gyges had access to extraordinary wealth, the kind that comes from controlling a river that carries gold in its current and a kingdom positioned at the crossroads of two civilizations. Second, he spent a remarkable amount of that wealth trying to secure legitimacy from the most respected religious authority in the Greek world. A secure king does not need to buy divine approval year after year, decade after decade. A king haunted by the manner of his ascension does, and the archaeological record suggests that the gifts never stopped flowing westward across the Aegean for the duration of Gyges' reign.

The offerings were the currency of a guilty conscience, or at least the currency of a man who understood that power acquired through murder requires constant reinforcement. Gyges needed the world to know that the gods approved of him. Every golden bowl placed at Delphi was a public statement that the Oracle had spoken and the new dynasty was ordained by Apollo himself.

The archaeological record supports the literary sources in broad outline. The royal burial mounds at Bin Tepe, a vast necropolis spread across the plain near Sardis, include tombs that scholars have associated with the Mermnad dynasty.[3] These are not modest graves scraped into the earth. They are monumental structures. The largest mound, traditionally attributed to Gyges' descendant Alyattes, is one of the biggest burial structures in the ancient world, rising from the flat terrain like a man-made hill visible for miles in every direction.[5] The Mermnad kings wanted to be remembered. They wanted the landscape itself to bear witness to their power. The dynasty that Gyges founded on the night he stepped out from behind a bedroom door built tombs that would last for millennia.

The excavations at Sardis also revealed something else about the Lydian kingdom during and after Gyges' reign. The gold-refining workshops near the Pactolus River produced a technological innovation that would reshape civilization itself. The process of separating gold from silver in naturally occurring electrum, which made possible the minting of coins with standardized and reliable precious metal content.[4] Before this innovation, a lump of electrum could contain anywhere from twenty to eighty percent gold, and the only way to know was to test it. Standardized coinage changed that equation entirely. Scholars debate the precise chronology, but many numismatists believe that the first true coins in human history were produced in Lydia during the Mermnad period.[4] The earliest surviving examples are small lumps of electrum stamped with a lion's head, no larger than a fingernail, held today in museum collections from London to Berlin to New York.

This is worth pausing on. The man who seized power through a bedroom assassination may have presided over the invention of money. The first coins were small enough to carry in a closed fist, easy to conceal, easy to pass from one hand to another without anyone else noticing the exchange. They could be spent without anyone knowing their origin. Unlike barter, which required two parties to meet face to face with goods of roughly equal value, coinage abstracted value into a portable anonymous form. A coin did not carry the identity of the person who had earned it. It did not reveal what labor or transaction had produced it. It was, in a very real sense, the first technology of financial invisibility.[4]

This connection between Gyges and the invention of anonymous value will resurface in a later chapter of this book, when we examine the modern systems of invisible money that allow wealth to move across borders without accountability or oversight. For now, it is enough to note the resonance. The man at the center of Western civilization's most famous thought experiment about invisibility also sits at the origin point of the most consequential tool ever devised for making value invisible. The ring and the coin share a common ancestor, and the kingdom of Lydia is where both stories begin.

But the story as told so far is not the version that changed history. Everything described in the preceding pages comes from Herodotus, who wrote his *Histories* in the fifth century BCE.[1] His account is vivid and psychologically complex, sympathetic to nearly everyone involved. Candaules is foolish but recognizably human in his foolishness. The queen is wronged and responds with terrifying logic. Gyges is coerced, reluctant, and trapped between two impossible options. The moral landscape of the Herodotean version is messy in exactly the way that real life is messy. Nobody acts in a vacuum. Everybody is constrained by the actions of others. The tragedy arises not from anyone's free choice but from a web of pressure, obligation, and desperation that makes the outcome feel inevitable from the moment Candaules opens his mouth.

Roughly a century after Herodotus, a philosopher in Athens rewrote the story. He kept the name Gyges. He changed everything else. In Book II of the *Republic,* Plato has the character Glaucon tell a story about a man named Gyges.[2] But in this version, Gyges is not a bodyguard. He is a shepherd. He is not serving in the palace of a king. He is tending sheep on a hillside in Lydia, living the most ordinary life imaginable. There is no Candaules, no queen, no bedroom intrigue, no coercion of any kind. The entire political and moral context of the Herodotean account has been stripped away like scaffolding from a finished building.

Plato's Gyges is going about his ordinary business when an earthquake splits the ground open before him. A great chasm appears in the earth, and the shepherd descends into it. In the depths he finds a hollow bronze horse, and inside the horse lies a corpse larger than any human body, wearing nothing but a golden ring on one finger.[2] Gyges takes the ring from the dead giant's hand. He climbs out of the earth and returns to his fellow shepherds. There, sitting among them at their regular meeting, he discovers by accident that the ring has a remarkable property. When he turns the bezel inward against his palm, he becomes invisible to everyone around him. When he turns it back outward, he reappears.[2]

The shepherd does not agonize over what to do with this power. He does not consult a priest or confess to his fellows or wrestle with the moral implications of what he now possesses. The discovery of invisibility leads immediately and without hesitation to action. He arranges to be included among the messengers sent to report to the king. He goes to the palace. He uses the ring to seduce the queen. He uses the ring to murder the king. He takes the throne.[2] The same destination as the Herodotean story. A completely different road to reach it.

In Plato's version, nobody forces Gyges to do anything. There is no queen issuing ultimatums. There is no king making perverse demands. There is no threat of death hanging over an impossible choice. Gyges discovers that he is invisible, and he immediately uses that invisibility to acquire sex, power, and a crown. The transition from shepherd to king happens through nothing but the exercise of free will enhanced by the certainty that no one can see what he is doing.

The philosopher's point is delivered through the character Glaucon, who uses the story to pose a question that has haunted moral philosophy ever since. Glaucon argues that any person, given a ring of invisibility, would behave the same way. The just man and the unjust man would become indistinguishable, because justice is merely the product of social pressure and the fear of punishment. Remove the possibility of being caught, Glaucon suggests, and the moral difference between people vanishes entirely.[2]

The question Glaucon puts to Socrates is this. If you could do anything without consequences, without anyone ever knowing, would you still choose to be good? Would you still refrain from taking what you wanted? Would justice survive the death of accountability? The rest of the *Republic* is, in many ways, Socrates' extended attempt to answer that question. His response spans ten books and encompasses a complete theory of justice, the human soul, and the ideal state. Socrates argues that the truly just person would not use the ring, that justice is its own reward, that the soul of the unjust person is deformed by injustice regardless of whether anyone sees the act. But the question itself has outlived the answer. Centuries later, people are still arguing about what we would really do with a ring that made us invisible.

What Plato did to the Gyges story was surgical and deliberate. He removed every element that made the historical Gyges sympathetic. In Herodotus, Gyges protests against the king's plan. He begs the queen for mercy. He acts under mortal threat with death as the only alternative. He is a man caught between forces larger than himself, a soldier trapped in the wreckage of his king's bad judgment. In Plato, the shepherd just... takes what he wants. The coercion is gone. The reluctance is gone. The web of obligation and fear is gone. What remains is pure, unforced choice exercised in the absence of any external pressure at all.[2]

This was not carelessness on Plato's part, and it was not a philosopher's ignorance of the historical tradition. Plato almost certainly knew the Herodotean version of the story. The *Histories* was one of the most widely read works in the Greek-speaking world by the time Plato was writing, and Plato was one of the most voraciously well-read men in Athens. The changes were deliberate. They had to be, because the moral thought experiment only works if the choice is free.

Consider why this matters. If Gyges is coerced, if he acts under threat of death with a queen's blade at his back, then his behavior tells us nothing about human nature in general. It tells us only that people prefer living to dying, which is hardly a revelation worth building a philosophy around. But if Gyges is free, if he finds a magic ring by accident and immediately uses it to commit adultery and murder without a moment of hesitation, then his behavior tells us something deeply uncomfortable about what people become when the social leash is removed. It suggests that morality is not an internal compass but an external cage, and that the person inside the cage would break free the instant the lock was opened.

Herodotus asked a historical question, what happened in Sardis around 680 BCE? Plato asked a philosophical one, what would happen anywhere, anytime, if the constraints on human behavior were removed entirely? The first question has a specific answer buried in the archaeological record, the literary sources, and the gold-bearing sands of the Pactolus. The second question has no definitive answer, which is precisely why it remains alive after centuries of debate.

The distinction between these two versions of the Gyges story is the engine that drives this entire book. Every chapter that follows will examine a real situation in which human beings found themselves unobserved, unaccountable, or anonymous. Some of those situations arose from technology that created new forms of invisibility. Some arose from institutional failures that left people without oversight. Some arose from the simple accident of being alone in the right place at the right time with no one watching. In each case, the question is the same one Glaucon posed in the *Republic*. What did these people do when no one could see them?

The answer, as we will see, is complicated. It is not a simple story of universal corruption in which every invisible person becomes a thief or a predator. Some people given the ring of invisibility act exactly as Glaucon predicted. They take what they can. They indulge appetites they would never reveal in daylight. They become unrecognizable versions of themselves, strangers to the people who thought they knew them. Others, given the same freedom from observation, do something entirely different. They become more generous, more honest, more authentically themselves than they ever manage to be when the world is watching. The ring does not create a single outcome. It creates a test, and different people pass or fail that test in ways that are surprising, disturbing, and occasionally inspiring.

But we are getting ahead of ourselves. The question is on the table now. The philosopher who asked it has been dead for centuries. The civilization that produced him turned to dust and was rebuilt and turned to dust again. The ring he described was never real. No golden band was ever pulled from the finger of a giant in a cave beneath the earth. No shepherd ever turned a bezel and vanished from sight. But the invisibility that ring represents, the ability to act without being seen, without being known, without being held accountable, has never been more available than it is right now.

References

[1] Herodotus. *The Histories*. Book I, Chapters 8-14. Translated by Robin Waterfield. Oxford University Press, 1998.

[2] Plato. *Republic*. Book II, 359a-360d. Translated by G.M.A. Grube, revised by C.D.C. Reeve. Hackett Publishing, 1992.

[3] Hanfmann, George M.A. *Sardis from Prehistoric to Roman Times*. Harvard University Press, 1983.

[4] Kraay, Colin M. *Archaic and Classical Greek Coins*. University of California Press, 1976.

[5] Roosevelt, Christopher H. *The Archaeology of Lydia, from Gyges to Alexander*. Cambridge University Press, 2009.

[6] Pedley, John Griffiths. *Sardis in the Age of Croesus*. University of Oklahoma Press, 1968.

Chapter Two - The Argument That Won't Die

The road from the Piraeus ran uphill toward Athens. It was a dusty path in the late afternoon light, cutting through olive groves and low stone walls. Somewhere around 375 BCE, a man named Socrates was walking that road after attending a festival at the harbor. He had watched the new torchlight procession for the goddess Bendis. He had said his prayers. Now he was heading home.

He did not make it very far. A group of young men spotted him from behind and sent a slave running to catch him. The slave tugged at his cloak and asked him to wait. Socrates turned around to find Polemarchus, the son of a wealthy arms manufacturer named Cephalus. Polemarchus was grinning. He had friends with him, including two brothers named Glaucon and Adeimantus. They wanted Socrates to come back to the house for dinner and conversation. "You could stay for the torch race on horseback tonight," Polemarchus offered. Socrates protested that he needed to get home, but the young men outnumbered him. "Could you persuade us to let you go, if we refuse to listen?" Polemarchus asked. The question was half a joke and half a philosophical trap.[1] Socrates relented. He walked back down the hill toward the harbor, toward a conversation that would last all night, across ten books of dialogue, through centuries of arguments. He could not have known what he was walking into. None of them could have.

The house of Cephalus sat near the Piraeus, comfortable and well-appointed. Cephalus himself was an old man by this point, a retired businessman who spent his remaining days thinking about the gods and whether he had lived a decent life. The conversation started casually enough. Socrates asked the old man about aging, about money, about what justice meant to someone who was running out of time to practice it. Cephalus gave a conventional answer... justice means telling the truth and paying your debts. Then he excused himself to attend to a sacrifice, and his son Polemarchus picked up the argument.[1] The talk moved in circles for a while. A combative Sophist named Thrasymachus barged into the discussion and declared that justice was simply whatever served the interests of the powerful. The strong make the rules, so "justice" is just another name for obedience to the strong. Socrates dismantled him methodically, exposing the contradictions in his position until Thrasymachus sat down in frustrated silence.[1] The evening might have ended there. The group might have moved on to wine and lighter subjects. But then Glaucon stood up. He was not satisfied.

The man who would pose the most dangerous question in the history of moral philosophy was not a professional philosopher. Glaucon was one of Plato's older brothers in real life. He was a young Athenian aristocrat who had likely served in the military. Ambitious and well-connected, he moved in political circles with designs on public life.[1] Sharp, restless, and unafraid of uncomfortable conclusions, he brought an intensity to conversation that few could match. Adeimantus, his brother, shared these qualities. These were real people living in a real city that had recently suffered devastating trauma. Athens had lost the Peloponnesian War less than a generation earlier. The great democracy had collapsed. A group of Athenian oligarchs known as the Thirty Tyrants had seized power under Spartan patronage and murdered an estimated fifteen hundred citizens in eight months of rule.[7] The democracy was eventually restored, but the scars remained. Everyone in that room at the house of Cephalus would have known people who were killed during the terror. Everyone would have known neighbors who collaborated with the Thirty, ordinary citizens who turned informant or looter once the old rules stopped applying.

Socrates himself was already a controversial figure in this period. Within a few years of this imagined conversation, the Athenian assembly would put him on trial for corrupting the youth and introducing strange gods. They would sentence him to death by a narrow margin. He would drink the hemlock in a prison cell surrounded by weeping friends, and Plato would spend the rest of his life haunted by it. The Republic was written partly as an act of mourning for his teacher and partly as a furious response to the city that killed him.[1],[7] The question of justice was not abstract to these men. They had watched their neighbors turn on each other. They had seen respectable citizens become tyrants overnight when the opportunity arose. The question of what people do when the rules stop working was something they knew in their bones.

Glaucon cleared his throat and addressed Socrates directly. He told him that Thrasymachus had given up too easily. The argument that justice is a sham had not been properly made, and therefore it had not been properly defeated. Glaucon wanted to hear the strongest possible version of the case against justice, stated with full intellectual force. Then he wanted Socrates to destroy it.[1]

He was explicit about his motives. Far from believing that injustice was better than justice, he wanted to believe the opposite. But he had heard this argument his whole life, in different forms and from different mouths, and nobody had ever beaten it convincingly. "I have never yet heard the superiority of justice to injustice maintained by anyone in a satisfactory way," he told Socrates. "I want to hear justice praised in respect of itself."[1] So he proposed to do something unusual. He would take the side he did not believe in, arguing for the superiority of injustice as forcefully as he could. He would make himself the devil's advocate and dare Socrates to prove him wrong. The fact that Glaucon wanted to lose was precisely what made the challenge so dangerous. He was not a cynic trying to score points. He was a sincere man constructing the strongest possible case against his own convictions, begging someone to show him where it failed.

The first argument landed like a stone thrown into still water. Glaucon proposed that justice is not natural to human beings. It is an invention born from weakness. People do not refrain from stealing because stealing is wrong. They refrain from stealing because they are afraid of being stolen from. The strong would prefer to take whatever they want, but even the strong cannot be strong all the time. Everyone is vulnerable when sleeping. Everyone can be outnumbered. So, people made an agreement, a compact of mutual restraint. They entered into what Glaucon called a social contract, though he did not use that exact phrase. The deal was simple... I will not harm you if you will not harm me. Justice, in this account, is a compromise between the best outcome (doing whatever you want without consequences) and the worst outcome (having terrible things done to you without recourse).[1] It is not a virtue. It is a truce. Nobody genuinely wants to be just. People settle for justice the way a traveler settles for a mediocre inn when the road has grown dark and dangerous.

The second argument was the thought experiment that would echo through the centuries. Glaucon told the story of a shepherd named Gyges who served the king of Lydia. Glaucon's point was not about magic rings. The ring was a device for isolating a single variable. The presence or absence of consequences. Suppose you gave one ring to the most just person alive and another to the most unjust. Then you watched what happened. Glaucon argued that both men would end up doing the same things. The just man would steal, cheat, sleep with anyone he wanted, free prisoners, kill anyone who stood in his way. "No one, it seems, would be so incorruptible that he would stay on the path of justice," Glaucon declared. "No one would keep his hands off other people's property."[1] The just person and the unjust person would become indistinguishable once the consequences were removed. Justice was never a property of the person. It was a property of the situation. Take away the situation and justice vanishes like smoke.

The third argument twisted the knife further. Glaucon asked Socrates to imagine two lives placed side by side for comparison. The first was the life of a perfectly unjust person who is so skilled at injustice that nobody ever catches him. This person lies, cheats, steals, and manipulates his way to the top while maintaining a flawless reputation for virtue. Everyone loves and respects him. He receives honors and praise. His children marry well. He gives lavish gifts to the temples and is celebrated as a pillar of the community.[1] The second life belonged to a perfectly just person who somehow receives a perfectly unjust reputation. This person never does anything wrong, but everyone believes he is a criminal, a fraud, a monster. He is stripped of his possessions, beaten and tortured. He is publicly humiliated and ultimately impaled.[1]

Now compare these two lives, Glaucon said. If justice has value "in itself," if being good is its own reward regardless of what the world thinks, then the just man on the rack should be happier than the unjust man on the throne. If that sounds absurd, then perhaps justice really is nothing more than a game of reputation. Perhaps the only thing people care about is the appearance of goodness, not goodness itself. The challenge was brutal in its clarity. Glaucon had separated the appearance of justice from the reality of it. He was demanding that Socrates defend the reality without any appeal to consequences, rewards, reputation, or the afterlife. He was asking a question that would reverberate for millennia. Is there any reason to be good when no one is watching and no punishment is coming?

The room fell silent. Adeimantus then rose to reinforce his brother's argument from a different angle. He pointed out that even the poets and the priests made justice sound like a transaction. The gods reward the just and punish the wicked, they promised. Parents tell their children to be good because goodness leads to prosperity. But all of this was still consequentialism. All of it still meant justice was only worth practicing because of what you got out of it. If you could fool the gods too... then what?[1] Adeimantus pushed even further. He noted that the religious traditions of their time actually offered rituals and sacrifices that could supposedly cleanse a person of injustice. You could sin all you wanted and then pay a priest to purify your soul. Even religion, it seemed, treated justice as something negotiable. Between the two brothers, they had constructed one of the most airtight philosophical challenges ever articulated. They laid it at the feet of Socrates and waited.

Socrates did not answer quickly. He acknowledged the weight of the challenge and told Glaucon plainly that the question was enormous. Then he said something that has frustrated readers for over two thousand years. He told Glaucon that the question was too complex to answer by looking at an individual person. They needed to look at something larger first. They needed to build a city.[1] What followed consumed the remaining eight books of the Republic. Socrates constructed an imaginary city from scratch, building it up class by class and institution by institution. His argument was that the same principles governing a just city would also govern a just soul. The city was a magnifying glass through which the soul could be examined at a scale large enough to understand. His core insight was structural, almost architectural. The soul, he argued, consists of three parts. The first is reason, the capacity for thought and truth-seeking. The second is spirit, the part that responds to challenges, that feels anger at injustice and pride in accomplishment. The third is appetite, the part that craves pleasure, food, drink, sex, money, and material comfort.[1]

Justice, Socrates claimed, is the proper ordering of these three parts.

When reason rules, with spirit as its loyal ally keeping appetite in check, the soul is harmonious. The just person is one whose inner life is well-governed. Each part performs its proper function without overstepping its boundaries. The unjust person, by contrast, is one whose appetites have overthrown reason and enslaved spirit. This inner tyrant makes the unjust person miserable regardless of external success. He might have wealth and power, but internally he lives in chaos, driven from desire to desire without any stable center.[1]

Socrates pressed the analogy further with an image that made the stakes vivid. Consider a city ruled by a tyrant, he said. From the outside it might look powerful. It might command armies, extract tribute, and silence all dissent. But inside its walls the citizens live in fear. Nobody trusts anyone. Friendships are impossible because everyone is an informant or a potential threat. The economy warps around the tyrant's personal appetites rather than serving the common good. The city is sick even when it appears strong.[1] The unjust soul operates the same way. A person driven by unchecked appetite might accumulate wealth and power, but his inner life is chaotic. He is enslaved by his own desires. His own judgment cannot be trusted because reason has been deposed. Genuine satisfaction eludes him because appetite always demands more. He is, in the deepest sense, unfree. The most powerful tyrant in the world is also the most enslaved person in the world, because he is a slave to himself.

Then Socrates offered the allegory of the cave, which has become perhaps the most famous image in all of Western philosophy. Imagine prisoners chained in an underground cave since birth, he told Glaucon. They sit in a row facing a blank wall. Behind them burns a fire, and between the fire and the prisoners, other people walk back and forth carrying objects above their heads. The prisoners can see only the shadows of these objects cast on the wall in front of them. They have never seen anything else. The shadows, to them, are reality. They give the shadows names and develop expertise in predicting which shadow will appear next. The best shadow-predictors are celebrated as wise men.[1]

Now imagine one prisoner is unchained and dragged up out of the cave into the sunlight. At first the light would blind him. It would hurt. He would want to go back to the comfortable darkness where everything made sense. But gradually his eyes would adjust. He would see the real world for the first time. Trees, rivers, mountains, the sun itself. He would realize that everything he had believed was a shadow of a shadow, a dim projection of a reality he had never suspected.[1] If this prisoner returned to the cave and tried to tell the others what he had seen, they would think he was insane. His eyes, now adjusted to sunlight, would be useless in the dark. He would stumble over things the other prisoners navigated easily. They would point to this as proof that his journey had damaged him. They might even try to kill him if he attempted to drag them up toward the light.[1]

The connection to Socrates' own fate was impossible to miss. But the philosophical point was about justice and knowledge. Most people, Socrates argued, mistake the shadows of social convention for reality. They think justice means appearing good, that happiness means satisfying appetites. They chase shadows on the wall and call it a meaningful life. The truly just person is the one who has climbed out of the cave, who has seen reality clearly, who understands the Good itself rather than its flickering imitations. Such a person would not be tempted by the ring because the ring offers nothing real. It offers only shadows. It offers the ability to manipulate a world of illusions, which is no prize at all for someone who has seen the sun.

It was a magnificent answer. It was also, in the eyes of many philosophers who came after, not entirely convincing.[1],[7] The problem was not the logic. The problem was the assumptions. Socrates' argument depended on accepting that the soul has exactly three parts, that reason is naturally the ruler among them, that knowledge of "the Good" is achievable in the way he described, and that inner harmony always trumps external suffering. If you accepted all of these premises, the conclusion followed beautifully. If you questioned any of them, the whole structure wobbled. Could a person not have a perfectly harmonious inner life organized around terrible values? Could reason not serve appetite rather than governing it? Was the metaphor of the cave anything more than a metaphor? These questions lingered. The man who raised them most effectively was someone who had studied in Plato's own school.

Aristotle arrived at Plato's Academy in Athens around 367 BCE. He was seventeen years old, a doctor's son from the northern frontier town of Stagira. He would stay at the Academy for twenty years, absorbing Plato's teachings before systematically dismantling many of them.[2] His disagreement with Plato on the nature of justice was not hostile. It was familial, the way a brilliant student corrects a beloved teacher by building something better on the same foundation.

Aristotle agreed with Socrates that the ring should not tempt a truly virtuous person. But he rejected the underlying theory of why. Socrates had argued that justice comes from knowledge. If you truly understand the Good, you will naturally do good. Injustice is merely ignorance. Know better and you will do better.[1] Aristotle thought this was wishful thinking. He had spent decades observing human behavior. He had watched intelligent people do foolish things and knowledgeable people act against their own stated principles. The gap between knowing and doing was enormous. Knowledge alone did not produce virtue.[2]

Virtue, Aristotle argued, is a habit. It is built through repeated practice within a community that values it. You become just by doing just things over and over, the way you become a carpenter by building furniture or a musician by playing scales. It is not an intellectual revelation that strikes like lightning. It is a craft that must be practiced daily. "We become just by doing just acts, temperate by doing temperate acts, brave by doing brave acts," he wrote in the Nicomachean Ethics.[2] The emphasis on repetition was deliberate and grounded in what he had observed of actual human development. A single courageous act does not make a person courageous. A lifetime of courageous acts does. The virtue must become second nature, as automatic as breathing, so deeply embedded in the character that it operates even under extreme pressure. The truly courageous soldier does not deliberate about whether to hold the line. He holds the line because that is who he has become.

What would happen to Aristotle's virtuous person if handed the Ring of Gyges? Aristotle believed such a person would not be tempted, because the desire to act justly would be so thoroughly ingrained that injustice would feel physically repulsive. It would be like asking a master musician to deliberately play out of tune. He possesses the technical ability to hit wrong notes, but everything in his trained sensibility rebels against it. The ring would feel wrong in his hands.[2] The problem, Aristotle freely admitted, was scarcity. Truly virtuous people are rare, far rarer than Socrates seemed to believe. Virtue requires the right upbringing in a well-ordered household. It requires the right community with strong institutions and good role models. It requires the right opportunities for practice across many different situations. It requires adequate material resources, because a starving person cannot cultivate temperance. It even requires a fair amount of luck, because a person born into poverty or tyranny may never encounter the conditions necessary for moral development.[2]

Most people never receive this complete moral education. Most people are somewhere in the middle, capable of justice when conditions support it and capable of injustice when conditions permit it. They are works in progress that never quite reach completion. Aristotle's answer to Glaucon was therefore conditional and honest in a way that Socrates' answer was not. The ring would not corrupt a virtuous person, but most people are not fully virtuous. The solution was not a better philosophical argument. The solution was better institutions, better education, better communities that shape character from childhood onward. The answer to the ring was not logic. It was formation, and formation requires a village that has decided, collectively, to raise its children well.

Let's explore the argument in more recent past...

Nearly two thousand years passed before anyone took Glaucon's side with full intellectual force. Thomas Hobbes was born in 1588, supposedly prematurely, because his mother was terrified by news of the approaching Spanish Armada. "My mother gave birth to twins... myself and fear," he reportedly said later in life.[3] Whether or not the story is literally true, it captures something essential about the man who told it. Fear was the lens through which Hobbes understood the world. He had good reason for it.

Hobbes lived his entire adult life in the shadow of political violence. He was a tutor to aristocratic families, a translator of Thucydides, a mathematician, and a political theorist who watched the English Civil War tear his country apart. He saw King Charles I put on trial by his own parliament and then beheaded on a scaffold in front of a crowd. Parliamentary armies swept through the countryside, burning manors and looting churches. Neighbors turned on neighbors over questions of religious doctrine that had seemed trivial just years before. He saw the entire structure of English society crack open, and he stared into what lay beneath.[3]

When Hobbes sat down to write Leviathan in 1651, he was not conducting an abstract exercise in political theory. He was trying to understand what he had witnessed. His conclusion was stark and unsparing. Human beings in their natural state, without government, without law, without enforceable agreements, exist in what he called "a war of all against all." Life without a sovereign authority is "solitary, poor, nasty, brutish, and short."[3] The phrase has become so famous that it has lost some of its original force. But Hobbes meant every word of it. He had watched it happen in real time.

Hobbes essentially agreed with Glaucon on every point. Justice is not natural. It is a contract born from fear. People obey the rules because they dread the consequences of disorder. Remove the consequences and the contract collapses. Give a man the Ring of Gyges and he will use it, because self-interest is the fundamental engine of human behavior. The desire to acquire more, to dominate, to satisfy appetite without restraint is built into our nature. Glaucon was right about all of it.[3]

Where Hobbes departed from Glaucon was in his conclusion. Glaucon had posed the ring as a test of individual virtue. Hobbes saw it as a problem of institutional design. The solution was not to make people better. The solution was to make the ring impossible. Hobbes called his ideal state the Leviathan, after the biblical sea monster, a sovereign authority so powerful that no individual could escape its reach. The Leviathan holds a monopoly on legitimate violence and uses that monopoly to enforce the social contract. People behave justly not because they are just, but because the Leviathan is always watching. The ring, in Hobbes' system, cannot exist. The sovereign sees everything.[3]

It was an honest answer, perhaps the most brutally honest answer anyone had given in the twenty centuries since Glaucon posed the question. It was also deeply pessimistic. If Hobbes was right, then justice is nothing more than a policing problem. Remove the police and the looting begins. The only thing separating civilization from savagery is the credible threat of punishment. Virtue is an illusion maintained at gunpoint. Every kindness, every act of generosity, every moment of compassion is underwritten by the implicit knowledge that someone with a sword is watching. Remove the watcher and the performance ends.

Immanuel Kant was born in 1724 in the Prussian city of Konigsberg, on the Baltic coast. He never traveled more than ten miles from his birthplace in his entire life. His daily walks were so precisely timed that neighbors reportedly set their clocks by his passing. He was a small, frail, extraordinarily disciplined man who lived almost entirely in the world of ideas, eating modestly, sleeping little, and spending his days in thought. He never married. Few close friends shared his solitary routine. He lived and died in the same city, in the same routines, producing philosophical works of such depth that they reshaped the intellectual landscape of the Western world.[4]

Kant was appalled by the implications of arguments like those of Hobbes. If morality depends on consequences, then morality is just self-interest in disguise. If you only tell the truth because lying might get you caught, then you are not being moral. You are being strategic. A person who refrains from stealing only because he fears prison is not virtuous. He is merely cautious. Kant wanted to establish that certain moral laws are absolute, unconditional, and binding on every rational being regardless of circumstance, consequence, or observation.[4]

He called his supreme principle the categorical imperative, and he formulated it in several ways. The most famous version runs roughly as follows. Act only according to rules that you could rationally will to become universal laws. Before performing any action, ask yourself what would happen if everyone in the world did exactly the same thing. If the result would be self-defeating or contradictory, the action is immoral. The test is logical, not emotional. It does not ask how you feel about an action. It asks whether the action is rationally coherent when universalized.[4]

Consider lying. If everyone lied all the time, language itself would cease to function. Communication depends on a baseline assumption of truthfulness. A world where lying is universal is a world where lying is pointless, because nobody believes anything anyone says. The very concept of deception requires a background of honesty against which it operates. Therefore, lying is categorically wrong. It is wrong even when the consequences of telling the truth are terrible. It is wrong even when no one would ever find out. It is wrong for the person wearing the Ring of Gyges just as surely as it is wrong for the person standing in full view of the public square.[4]

Kant's response to Glaucon was the philosophical equivalent of a door slamming shut. The ring is irrelevant. Morality does not come from observation, reputation, reward, or punishment. Morality comes from reason itself. A rational being recognizes the categorical imperative the way a mathematician recognizes a valid proof. You do not need an audience to know that two plus two equals four, and you do not need an audience to know that stealing is wrong. The moral law, Kant wrote, fills the mind with "ever new and increasing admiration and awe" when one reflects on it seriously. It stands alongside the starry sky above as evidence of something transcendent in the human condition.[4]

The strength of Kant's position was its purity. The weakness was the same. Kant described what morality should look like for a perfectly rational being. He did not describe what morality actually looks like for real human beings, who are rarely perfectly rational and who frequently know the right thing to do while doing the wrong thing instead. Every person who has ever eaten a second piece of cake while knowing they should not, every person who has snapped at a loved one while understanding that patience was the correct response, every person who has told a white lie to avoid an uncomfortable conversation understands the gap between moral knowledge and moral action. Kant could prove that the Ring of Gyges should not change a rational person's behavior. He could not prove that it would not.[4] The gap between "should" and "would" was exactly the gap Glaucon had identified twenty-two centuries earlier. Kant had not closed it so much as declared it philosophically unimportant. For anyone interested in what people actually do rather than what they ought to do, that declaration rang hollow.

Friedrich Nietzsche read Plato's Republic as a young man and reacted to it the way a match reacts to dry kindling. He was born in 1844 in the Saxon town of Rocken, the son of a Lutheran pastor who died when Nietzsche was four years old. He grew up brilliant, sickly, and profoundly alienated from the pious conventional world around him. By his late twenties he was a professor of philology at the University of Basel in Switzerland, the youngest person ever appointed to that position. By his mid-forties he had gone insane, collapsing on a street in Turin in January 1889. He would spend the final eleven years of his life in the care of his mother and then his sister, unable to write or think coherently.[5] Between those two bookends he produced some of the most incendiary philosophical writing in the Western tradition.

Nietzsche did not try to refute Glaucon's challenge. He embraced it and took it further than Glaucon himself had dared. The Ring of Gyges, Nietzsche argued, does not reveal that people are secretly immoral. It reveals that "morality" itself is a fiction created by the weak to enslave the strong.[5] The ring does not strip away virtue to expose vice. It strips away a collective illusion to expose reality.

Consider the origin of moral concepts, Nietzsche wrote in On the

Genealogy of Morality. In the ancient world, the strong, the noble, and the powerful originally defined "good" as whatever they themselves embodied. "Good" meant aristocratic, vital, commanding, beautiful, and free. "Bad" meant common, weak, low, and slavish. This was what Nietzsche called "master morality," and it was the original form of moral thinking.[6] The weak could not defeat the strong through physical force, so they defeated them through a revolution in ideas. They invented a new moral framework, one that Nietzsche called "slave morality," in which the meek were righteous and the strong were sinful. Humility became a virtue where once it had been a mark of servility. Pride became a vice where once it had been a mark of excellence. Submission was called holy. Ambition was called wicked. The entire moral vocabulary was inverted.[6]

Nietzsche saw this inversion as a great act of collective revenge disguised as ethics. Slave morality was the weapon the weak used to chain the strong, and it worked precisely because it convinced everyone, including the strong themselves, that the chains were sacred. Glaucon's argument was therefore not a problem to be solved but a truth to be acknowledged. The social contract does constrain the strong on behalf of the weak. Conventional morality is a cage. The Ring of Gyges simply opens the cage door.[5]

The figure Nietzsche proposed as the answer to this situation was the Ubermensch, the "overman" or "beyond-man." This was not a figure of crude brute power, though many later readers would misunderstand him that way. The Ubermensch was someone who created his own values rather than passively accepting the inherited moral framework of the herd. He would use the ring without guilt because conventional morality had no legitimate claim on him. He had moved beyond good and evil into a realm where he answered only to the standard he had forged for himself.[5],[6]

Nietzsche's position was exhilarating and terrifying in equal measure. It was exhilarating because it named something that more polite philosophers refused to acknowledge. Much of conventional morality does function as a tool of social control. The powerful do frequently use the language of virtue to justify their dominance while punishing anyone who threatens their position. The history of religious institutions alone provides centuries of evidence for this claim. Nietzsche cut through the hypocrisy with surgical precision, forcing his readers to confront uncomfortable truths about the origins and functions of the moral systems they took for granted.

It was terrifying because the logic led to places Nietzsche himself may not have fully anticipated. If morality is a fiction and the strong should create their own values, who decides what "strong" looks like? Who determines which self-created values are noble and which are merely destructive? In the twentieth century, people who misread Nietzsche and use his vocabulary to justify horrors he might have found repulsive. The Ubermensch was supposed to be an artist of the self, a creative genius who transcended petty conventions. He became, in lesser and more brutal hands, an excuse for monsters.

Nietzsche did not refute Glaucon. He celebrated him. He argued that the discomfort people feel when confronting the Ring of Gyges is not evidence that Glaucon was wrong. It is evidence that slave morality has done its work so thoroughly that most people cannot even imagine an alternative. The chain has become invisible to those who wear it. The prisoner has come to believe the cage is the world.

Centuries after Glaucon stood up in the house of Cephalus, his argument remains undefeated. That is not to say the responses have been weak. Each of the great thinkers who grappled with the ring brought genuine insight to the question. Socrates identified something important about the relationship between internal harmony and external behavior. A person in a state of inner chaos is unlikely to experience genuine happiness regardless of material success. This observation has held up surprisingly well in the age of modern psychology, where research consistently links internal coherence to wellbeing. Aristotle was right that character is built through habit and that the environment in which a person is raised profoundly shapes their moral capacities. Developmental psychology has confirmed this insight in detail he could never have imagined. Kant was correct that moral reasoning should aspire to universality, that a world where everyone lies is a world that cannot function. Game theorists have proven this mathematically. Even Nietzsche, for all his dangerous implications, was honest about the ways morality can serve as a mask for power in ways that the other philosophers preferred not to examine.

The trouble is that each response addresses a different aspect of the question while leaving the central challenge intact. Socrates argues that the ring should not matter because the just soul is inherently happier. But this requires accepting a specific metaphysical picture of the soul that many people find unpersuasive.[1],[7]

Aristotle argues that the virtuous person would resist the ring because virtue has become habitual. But he concedes that most people are not fully virtuous and therefore would not resist.[2]Kant argues that the ring is morally irrelevant because duty transcends consequences. But he cannot explain why so many apparently rational beings fail to obey the categorical imperative even without the ring, even in broad daylight, even when they know they are being watched.[4]

Hobbes and Nietzsche, who agree with Glaucon that the ring reveals human nature as it truly is, lead us to conclusions that make civilization itself seem like an elaborate bluff, a thin shell of agreements stretched over a molten core of appetite and self-interest.[3],[5]

The relay race of brilliant minds has been running for millennia, and no one has crossed the finish line. The baton passes from century to century, globally and the fundamental question stays open. Is justice real, or is it merely performed? Is goodness a feature of the soul, or is it a feature of the situation? If you could act without consequences, would you still choose to be good?

The reason the argument will not die is that it is, at its core, an empirical question masquerading as a philosophical one. Socrates, Kant, and Aristotle tried to answer it through logic, through reason, through the architecture of careful argument. Their answers were elegant and internally consistent. But their answers were also untestable. You cannot hand someone an actual Ring of Gyges and observe what happens. You cannot run the experiment under controlled conditions. You cannot collect data. For most of human history, the question belonged to philosophy precisely because it could not belong to science. There was no way to watch millions of people making moral decisions while believing they were unobserved. There was no way to study what happens when an entire society is given the functional equivalent of Glaucon's ring. The argument survived because the evidence was always somewhere beyond the horizon, tantalizingly out of reach.

Plato wrote the Republic in the fourth century before the common era. Aristotle taught at the Lyceum as Alexander the Great conquered the known world. Hobbes watched the English monarchy collapse and then rebuild itself from the wreckage. Kant walked his precise daily route through Konigsberg while empires rose and fell around him. Nietzsche went mad in Turin in January of 1889, collapsing in the street while reportedly embracing a horse that was being beaten by its owner. Each of them lived and died in a world where the Ring of Gyges was purely hypothetical, a thought experiment that could never become a real experiment. None of them could have imagined a technology that would give ordinary people the ability to act anonymously on a global scale, to adopt false identities with a few keystrokes, to manipulate others without ever revealing themselves, to commit acts that left no visible trace and carried no immediate consequence. None of them could have imagined the internet.

The thought experiment that Glaucon proposed in a torchlit dining room near the harbor of Athens is no longer a thought experiment. It has been running, in real time, across the entire connected world, for more than three decades now. Billions of people have been handed a version of the ring. They did not find it in a chasm after an earthquake. They bought it at an electronics store or received it as a gift on a birthday or found it waiting for them in a school computer lab. It rides in their pockets. It sits on their nightstands. They stare into its glow before falling asleep.

The question of what they do with it is no longer philosophical. It is documented. It is measured. It is stored on servers in data centers from Virginia to Singapore. The argument that would not die for centuries has finally collided with evidence. And the evidence... is complicated.

References

[1] Plato, *Republic*, trans. G.M.A. Grube, rev. C.D.C. Reeve (Indianapolis: Hackett, 1992). All quotations from and references to the dialogue between Socrates, Glaucon, Adeimantus, and Thrasymachus are drawn from this translation unless otherwise noted.

[2] Aristotle, *Nicomachean Ethics*, trans. Terence Irwin (Indianapolis: Hackett, 1999). The discussion of virtue as habit, the comparison of moral development to craft training, and the conditions required for virtuous character are drawn from Books II and X.

[3] Thomas Hobbes, *Leviathan*, ed. Richard Tuck (Cambridge: Cambridge University Press, 1996). Originally published 1651. The "war of all against all" and "solitary, poor, nasty, brutish, and short" appear in Part I, Chapter 13.

[4] Immanuel Kant, *Groundwork of the Metaphysics of Morals*, trans. Mary Gregor (Cambridge: Cambridge University Press, 1998). Originally published 1785. The categorical imperative and the passage on "admiration and awe" are drawn from the Second and Third Sections.

[5] Friedrich Nietzsche, *Beyond Good and Evil*, trans. Walter Kaufmann (New York: Vintage, 1966). Originally published 1886. Nietzsche's critique of conventional morality and his concept of the Ubermensch are developed across this work and in *Thus Spoke Zarathustra*.

[6] Friedrich Nietzsche, *On the Genealogy of Morality*, trans. Maudemarie Clark and Alan J. Swensen (Indianapolis: Hackett, 1998). Originally published 1887. The distinction between "master morality" and "slave morality" is developed in the First Essay.

[7] On the Thirty Tyrants and the political context of late fifth-century Athens, see Xenophon, *Hellenica*, Book II; and Peter Krentz, *The Thirty at Athens* (Ithaca: Cornell University Press, 1982). The estimate of fifteen hundred citizens killed appears in Aristotle, *Athenian Constitution*, 35.4. Commentary on the Republic as a response to Socrates' trial and execution draws on several secondary sources, including C.D.C. Reeve, *Philosopher-Kings: The Argument of Plato's Republic* (Princeton: Princeton University Press, 1988).

Chapter Three - "On the Internet, Nobody Knows You're a Dog"

The cartoon is simple. Two dogs sit in front of a desktop computer. One dog occupies the chair. The other sits on the floor beside him. The seated dog looks down at his companion with the easy confidence of someone sharing a secret. The caption reads, "On the Internet, nobody knows you're a dog."

Peter Steiner drew it for *The New Yorker* in the summer of 1993.[1] He later admitted that he did not think much of it at the time. The cartoon was one of many he had submitted that year. He considered it a minor work, a throwaway gag about a technology he barely understood. The magazine ran it on page 61 of the July 5 issue, tucked between articles that no one remembers. Steiner collected his standard fee and moved on to the next batch of sketches.

The cartoon did not move on. It became the most reproduced cartoon in the history of *The New Yorker*, reprinted on coffee mugs and T-shirts and in the opening slides of countless academic presentations about digital culture. College professors photocopied it for their syllabi. Technology journalists quoted the caption as shorthand for everything the internet promised. For years, it circulated as a celebration. The internet was a place where your body did not matter. Your race did not matter. Your age, your appearance and your disability did not matter. You could be anyone. You could finally be judged by the quality of your ideas rather than the circumstances of your birth.

That was the reading in 1993. It was optimistic and generous, consistent with the spirit of the moment. The internet was still new enough to feel like a frontier, a territory where the old rules had not yet arrived. The cartoon captured the thrill of that possibility. You could slip free of every category that the physical world had forced upon you. Nobody knows you're a dog. Nobody knows you're shy. Nobody knows you're poor, old or stuck in a small town with small minds. The ring was invisible, and so were you. But there was another reading. It was present from the beginning, hiding in plain sight within the same six words. Nobody knows you're a dog. Nobody knows who you are, or what you are capable of. No one is watching. The cartoon was both a celebration and a warning. It took the world two decades to understand which reading would prevail.

Before the cartoon, before the World Wide Web itself, the networks existed in a form that their creators understood as something close to paradise. The story begins in the late 1960s with ARPANET, the United States Department of Defense project that linked a handful of university computers into a shared network.[3] The original nodes connected UCLA, Stanford Research Institute, UC Santa Barbara, and the University of Utah. The network's designers were physicists and engineers, people who solved problems by building systems. They built ARPANET to share computing resources across distances. They did not set out to invent a new form of human communication. That part was an accident.

The accident happened quickly. Within months of the first successful connection in October 1969, the researchers discovered that they preferred using the network to talk to each other rather than to share processing power. Email emerged before anyone had planned for it. Discussion groups followed. The network that was supposed to move data began moving conversations instead, and those conversations had a quality that felt different from anything these researchers had experienced before.

The difference was anonymity. Not total anonymity, since these early users knew each other by name and institution. But the communication itself stripped away the markers that shaped every face-to-face interaction. On a terminal screen, a graduate student's argument carried the same visual weight as a tenured professor's. The words arrived in the same monospaced font on the same green-black display. There was no oak-paneled office to intimidate, no tweed jacket to signal authority, no gray hair to demand deference. The hierarchy of the physical world did not translate through the wire.

This accident became an ideology. Through the 1970s and into the 1980s, as the network expanded from a few dozen nodes to thousands of connected systems, the culture that grew around it developed a set of shared beliefs that bordered on the religious. The hacker ethic, as the journalist Steven Levy later codified it, held that information should be free, that access to computers should be unlimited, and that authority should be judged by results rather than credentials.[3] These principles did not emerge from philosophy departments. They grew from the lived experience of people who had discovered that removing physical identity from conversation seemed to make conversation better.

Usenet arrived in 1980, a distributed discussion system that organized conversations into newsgroups arranged by topic. Unlike ARPANET, which required institutional access, Usenet spread across universities through ordinary phone lines using the UUCP protocol. The network grew from two connected sites to thousands within a few years. It was chaotic and unmoderated with pseudonymous by default. Users chose handles that revealed as much or as little as they wished. Discussions ranged from computer science to cooking to politics to pornography. The signal-to-noise ratio varied wildly, but the committed users believed they had found something new in human experience. They believed they had found a place where ideas could stand on their own merits.

The most articulate expression of this belief came from the Whole Earth 'Lectronic Link, known as the WELL. Stewart Brand and Larry Brilliant founded the WELL in 1985, building it on the infrastructure of Brand's Whole Earth Catalog counterculture.[4] The platform was a dial-up bulletin board system based in Sausalito, California. It charged a modest monthly fee. It attracted writers, hackers and Grateful Dead fans. It was a self-selected community of several thousand users who understood themselves as pioneers.

Howard Rheingold, one of the WELL's most prominent early members, captured the community's self-image in his 1993 book *The Virtual Community*. He described the WELL as a place where "people in virtual communities do just about everything people do in real life, but we leave our bodies behind."[4] The phrasing was deliberate. Leaving bodies behind meant leaving behind the prejudices attached to bodies. It meant creating a space where a person's contribution to the conversation was the only thing that mattered.

The WELL operated on a principle that its members called "You Own Your Own Words." Every post was attributed to a handle, and that handle accumulated a reputation over time. The system was pseudonymous rather than anonymous. Regulars knew each other. Trust built slowly through thousands of small exchanges. The community developed norms and enforced them through social pressure rather than technological control. When someone violated the spirit of the community, other members would respond. The system worked because it was small enough for everyone to matter and persistent enough for reputation to carry consequences.

For a brief window in the late 1980s and early 1990s, the WELL seemed to prove that the utopians were right. Strip away the body. Let the words speak for themselves. Trust the community to self-regulate. Human nature, freed from the distortions of physical prejudice, would trend toward generosity and honesty and mutual respect.

The prophet of this vision was a retired Wyoming cattle rancher named John Perry Barlow. Barlow was an unlikely figure for the role. He had written lyrics for the Grateful Dead and run his family's ranch in Pinedale, Wyoming. He had no formal training in computer science or philosophy. But he possessed a gift for translating technical possibilities into moral arguments, and in the early 1990s he became the most eloquent voice for the idea that the internet represented a fundamental break in human history.

He co-founded the Electronic Frontier Foundation in 1990 alongside Mitch Kapor, the creator of Lotus 1-2-3. The organization's name was deliberate. The internet was a frontier, and frontiers belonged to the brave and the free. Barlow spent the next several years arguing in essays and lectures that cyberspace was a new kind of territory, one that existing governments had neither the right nor the ability to control.

His most famous articulation came on February 8, 1996. Barlow was in Davos, Switzerland, attending the World Economic Forum. The previous day, President Bill Clinton had signed the Telecommunications Act of 1996 into law. The act included the Communications Decency Act, which attempted to regulate indecent material on the internet. Barlow viewed the legislation as an act of aggression by a world that did not understand what it was trying to govern. He sat down at a laptop in the conference center and composed a document that he titled "A Declaration of the Independence of Cyberspace."[2]

The declaration was addressed to "Governments of the Industrial World, you weary giants of flesh and steel." Its language was deliberately Jeffersonian. Barlow wrote, "We are creating a world that all may enter without privilege or prejudice accorded by race, economic power, military force, or station of birth."[2] He continued, "We are creating a world where anyone, anywhere may express his or her beliefs, no matter how singular, without fear of being coerced into silence or conformity."[2]

Barlow sent the declaration to approximately six hundred people on his email list. Within days, it had been forwarded to millions. It appeared on tens of thousands of websites. It was translated into dozens of languages. For a generation of internet users, it became a founding document, a statement of principles for the world they believed they were building.

The declaration's central argument rested on an assumption about human nature. Barlow believed that people who were freed from the constraints of the physical world would behave better, not worse. He believed that prejudice and cruelty were symptoms of physical scarcity. Bodies competed for territory. Bodies feared other bodies. But in cyberspace, where territory was infinite and bodies were absent, the ancient competitions that drove human conflict would lose their power. Remove the body, and you remove the motive for cruelty.

It was a beautiful argument. It was also the same bet that Glaucon described in the Republic, the same wager that Plato's Socrates spent an entire dialogue attempting to answer. Barlow was betting that the soul, freed from constraint, would choose justice. He was betting against the ring.

The cracks appeared almost immediately. In November 1994, a twenty-year-old University of Michigan student named Abraham Jacob Alkhabaz, who used the name Jake Baker, posted a short story to the Usenet newsgroup alt.sex.stories. The story described the kidnapping, torture, rape, and murder of a woman. Baker used the real name of a classmate.[5]

The story was graphic. It was also clearly labeled as fiction. Baker had posted similar material before under his Usenet handle. The newsgroup where he published it was dedicated to erotic fiction, and graphic violence was not uncommon in its archives. What made Baker's story different was the use of a real person's name. The woman he named was a student who sat near him in a Japanese language class. She did not know him. She did not know that he had written about her. She learned about it when a University of Michigan alumnus in Moscow read the story and contacted the university's administration.

The story attracted attention within the newsgroup. Several anonymous readers responded with encouragement. One of them, using the handle "Arthur Gonda," began exchanging private emails with Baker. In these emails, Baker and Gonda discussed plans to abduct and harm a woman. The emails were more explicit than the fiction. They read less like fantasy and more like logistics.

The FBI arrested Baker on February 9, 1995. Federal prosecutors charged him under a statute prohibiting the transmission of threats across state lines. The case became a national story, one of the first instances where the collision between online expression and real-world harm demanded legal adjudication. Baker's attorneys argued that the posts were fiction. The prosecution argued that the emails revealed genuine intent.

Judge Avern Cohn of the Eastern District of Michigan dismissed the charges.[5] The ruling held that Baker's posts, however disturbing, were protected speech. The emails with Gonda, while more troubling, did not constitute a "true threat" under existing legal standards. Baker was released. He returned to campus briefly before withdrawing from the university.

The legal resolution of United States v. Baker settled one narrow question about the boundaries of online speech. But the case established a pattern that would repeat itself thousands of times in the years that followed. The pattern had three elements. An anonymous or pseudonymous space removed the social consequences of expression. The absence of consequences enabled escalation. The audience, equally anonymous, provided encouragement that amplified the escalation further. Anonymity created permission. Permission attracted an audience. The audience rewarded the behavior. The cycle fed itself.

Baker did not need the internet to have violent fantasies. He needed the internet to share them under conditions where sharing carried no social cost. He needed the invisible readers of alt.sex.stories to respond with interest rather than alarm. What he also needed was the pseudonymous Arthur Gonda to match his escalation with equal enthusiasm. The ring did not create the impulse. The ring created the conditions under which the impulse could grow unchecked.

The Baker case was dramatic enough to attract national headlines. But the more revealing evidence came from ordinary conversations. Mike Godwin was a lawyer and one of the earliest legal thinkers to take internet culture seriously. In 1990, he made an observation that he phrased as a law of online discourse. "As an online discussion grows longer," Godwin wrote, "the probability of a comparison involving Nazis or Hitler approaches one."[6]

Godwin's Law was presented as humor. It circulated through Usenet as a joke, a knowing reference to the tendency of online arguments to lose proportion. Someone would post a political opinion. Another user would disagree. The disagreement would sharpen. Within a dozen exchanges, one participant would compare the other to Hitler. The thread would collapse into recrimination and insult. The pattern was so reliable that Godwin could describe it as a mathematical certainty.

The humor obscured the significance. Godwin's Law described a predictable degradation that occurred when human beings argued without faces. In a physical conversation, social feedback mechanisms constrain escalation. You can see when your interlocutor flinches. You can hear the tone shift in a room when you have gone too far. These signals are not moral arguments. They are bodily information, data transmitted through facial expression and posture and the subtle modulations of voice that human beings have spent millions of years learning to read. Online communication eliminated all of it.

The elimination was not incidental. It was structural. Text on a screen carries no facial expression. It conveys no tone of voice. It provides no information about the physical vulnerability of the person on the other end of the connection. The words arrive stripped of every contextual signal that might trigger empathy or restraint. The reader fills the silence with projection, and projection tends toward the adversarial. When you cannot see the person you are arguing with, it becomes easier to imagine them as contemptible. When you cannot hear them wince, it becomes easier to wound them.

Usenet's flame wars were the earliest laboratory for this dynamic. The newsgroups that hosted political discussion became notorious for the speed and depth of their degradation. Conversations that began with substantive disagreements about policy would devolve into personal attacks within hours. Users who had never met would develop bitter vendettas that persisted for months. The arguments generated more heat than light because heat was what the medium rewarded. A measured response disappeared into the scroll of new messages. An outrageous provocation generated dozens of replies. The structure of the platform selected for extremity.

The early trolls understood this instinctively. Trolling began as a form of play, a test of the community's ability to distinguish genuine participants from provocateurs. The term itself came from fishing, where trolling means dragging a baited line through the water to see what bites. On Usenet, a troll was a deliberately provocative post designed to elicit outraged responses. The troll succeeded when serious users took the bait.

In its earliest form, trolling was relatively harmless. A user might post an absurd claim to see how many people would respond with earnest corrections. The humor was in the gap between the troll's indifference and the respondent's sincerity. It was a game that required a certain intelligence. The best trolls crafted posts that were just plausible enough to draw responses but absurd enough to signal their true nature to the perceptive.

The shift from play to cruelty happened gradually. As the anonymous spaces of the internet expanded, trolling evolved from a game into a culture, and the culture developed its own logic of escalation. The early trolls had pranksters' ethics. They wanted to provoke laughter. The later trolls wanted to provoke suffering. They targeted specific individuals with coordinated campaigns of harassment. They used the anonymity of the medium not to test the community but to assault the vulnerable. The ring had always offered the choice between mischief and malice. As the population grew, more users chose malice.

Even the WELL, the community that the utopians had held up as proof of their thesis, discovered the limits of its own model.

The crisis came not from outsiders but from within. Members who had known each other for years through their handles discovered that pseudonymity enabled forms of deception that the community's norms could not prevent. Some members maintained multiple accounts, using different handles to manipulate discussions or to create the illusion of consensus. Others used the intimacy of the community's private messaging system to cultivate relationships built on fabricated identities. The violations were not criminal. They were social. They were betrayals of trust made possible by the gap between the online persona and the physical person.

Rheingold documented some of these tensions in his writing about the WELL. He acknowledged that the community he celebrated had experienced "weights and incremental pressure, the gradual revelation that other people's intentions are not always benign."[4] The admission was careful. Rheingold still believed in the model. He still believed that the benefits of online community outweighed the risks. But he recognized that the community he had described as a model for the future contained the same capacity for manipulation that existed in every human community. The technology hadn't changed human nature. It had only changed the tools available for expressing it.

The WELL's struggles were instructive because the community represented the best possible case for the utopian argument. Its members were self-selected. They were literate and educated, sharing cultural references and paying for access, which filtered out the casual and the uncommitted. Over years of shared experience, they had developed elaborate social norms. If anonymity and pseudonymity were going to produce a better form of human interaction, the WELL was where it should have worked.

It worked, for a while, imperfectly. The imperfections were not catastrophic. The WELL did not collapse. It survived and evolved and still exists in diminished form today. But the community's experience demonstrated that even small groups of well-intentioned people could not fully overcome the temptations that the ring presented. The promise that removing bodies from conversation would elevate that conversation proved true in some cases and false in others. The truth depended on the individuals involved, which was precisely the point that Plato had been making for centuries.

The decisive test came in September 1993, the same month that Peter Steiner's cartoon was still circulating through *The New Yorker*'s readership. For more than a decade, Usenet had operated as a community of communities. The network had no central authority. It had no formal moderation system. It had only the cultural norms that its users had developed through years of shared practice. New users arrived slowly, one at a time, through university accounts or through the handful of internet service providers that offered Usenet access. They learned the culture by observation, lurking before they posted. Gradually they absorbed the conventions of the newsgroups they joined. The process was gradual, and the gradualness was the mechanism by which the culture reproduced itself.

In September 1993, America Online opened a gateway that gave its millions of subscribers direct access to Usenet.[3] The influx was immediate and overwhelming. Thousands of new users arrived each day. They had not lurked. They had not absorbed the norms or learned the conventions. Their posts asked questions that the regulars considered elementary. Many cross-posted to inappropriate newsgroups. Others replied to trolls with earnest indignation. They treated the space as they treated every other consumer product they had purchased, which was to say they expected it to serve their needs without requiring them to adapt their behavior.

The existing Usenet community called it Eternal September. The name referenced the annual influx of new university students who arrived on Usenet each September when the academic year began. In previous years, these newcomers had been absorbed into the culture within a few weeks. By October, they had learned the rules. By November, they were indistinguishable from the regulars. But the AOL influx had no October. The newcomers never stopped arriving. The culture of Usenet, which had depended on a gradual process of socialization, could not scale to accommodate the flood.

Eternal September was the moment when the internet stopped being a community and became a population. The distinction matters. A community shares norms. Its members understand the unwritten rules because they have absorbed them through participation. A population shares only a platform. Its members bring their own norms, or no norms at all, and the platform itself provides no mechanism for reconciling the differences.

The pattern that Eternal September revealed would repeat itself across every subsequent platform that the internet produced. Small communities could self-regulate because their members were visible to each other, because reputation carried weight, because the social cost of violation was high enough to deter most violations. Large communities could not self-regulate because the mechanisms of social pressure that worked at small scale collapsed at large scale. When your audience is a dozen people who know you by your handle and remember what you said last month, you moderate your behavior. When your audience is ten million strangers who will never see your username again, the ring is absolute.

The early internet pioneers understood this problem. They discussed it at length. They proposed solutions, including moderation systems, reputation scoring, identity verification, and technological architectures that might preserve the benefits of pseudonymity while limiting its harm. Some of these solutions worked in limited contexts. None of them solved the fundamental problem, which was that the benefits of anonymity to the speaker came at a cost to the community. The speaker gained freedom. The community lost trust. And the mathematics of this exchange grew worse as the numbers increased.

The utopian vision had rested on a miscalculation about human nature. The early internet pioneers had assumed that people were fundamentally good and that the cruelties of the physical world were products of physical scarcity and physical prejudice. Remove the scarcity. Remove the prejudice. Give people a space where their words could speak for themselves. The result would be liberation.

The ring suggested otherwise. The ring's lesson, as Plato had articulated it, was that human behavior is shaped by observation and consequence. Remove the observer. Remove the consequence. The behavior changes. It does not change uniformly. Some people remain just without the ring. Some people discover appetites they did not know they had. Most people fall somewhere in between, capable of both generosity and cruelty, their choice determined by the specific conditions of the moment. The early internet amplified all three tendencies, but the architecture of scale, anonymity, and audience gave the cruelest voices the greatest reach.

John Perry Barlow lived long enough to see what became of his declaration. In interviews during the 2000s and 2010s, he addressed the question directly. He did not recant, maintaining that the principles of his declaration were correct even as he acknowledged that the internet had not produced the world he had envisioned. Instead, he drew a distinction between the principles and the practice. The principle that speech should be free remained valid. The practice of free speech on platforms designed to maximize engagement had produced something he had not anticipated.

The tension he wrestled with was the same tension that animated Plato's dialogue. Freedom enables the dissident and the predator equally. The same anonymity that protects a Chinese democracy activist protects a child predator. The same platform that gives voice to the marginalized gives a megaphone to the hateful. Barlow understood this. He had always understood it in the abstract. What surprised him was the ratio. He had expected that the dissidents would outnumber the predators. He had expected that the liberating uses of anonymity would predominate over the destructive ones. The mathematics did not work out that way.

Barlow's faith was not naive. He had grown up on a ranch in Wyoming. He understood that human beings were capable of cruelty. He understood that freedom carried risks. His argument had never been that all people would use the internet wisely. His argument was that the alternative to freedom was worse. The alternative was control, and control meant trusting some authority to decide which speech was acceptable. Barlow had looked at the governments of the world and concluded that none of them deserved that trust.

The argument held a certain logic. It also contained a blind spot that the ring illuminated. Barlow had framed the question as a binary between freedom and control. The ring suggested a third possibility. The possibility was that human beings in large numbers, given anonymity and an audience, would generate a form of discourse that was neither free nor controlled but simply degraded. Not censored, but loud. Not suppressed, but incoherent. Not silenced, but drowned in noise. The worst speech would not be banned. It would simply outcompete the best speech because sadly... outrage travels faster than nuance and cruelty generates more engagement than kindness.

Barlow died on February 7, 2018, at the age of seventy. He died in his sleep at his home in San Francisco. The internet he had championed mourned him with the tools he had helped make possible. Tributes appeared on Twitter and Facebook, platforms that had not existed when he wrote his declaration. The tributes shared space in the same feeds with harassment campaigns, conspiracy theories, and the routine cruelties of anonymous discourse. The juxtaposition was the point. The internet had not sorted itself into the utopia he had envisioned or the dystopia his critics feared. It had become both simultaneously, a space where liberation and degradation occupied the same infrastructure.

Steiner's cartoon hangs in the hallways of technology companies and communication departments at universities around the world. Its meaning has shifted over the decades, from celebration to warning to something closer to an epitaph.

- On the internet, nobody knows you're a dog.
- On the internet, nobody knows who you really are.
- On the internet, the constraints that maintained the uneasy peace of physical society dissolve into the glow of the screen.

The internet is the largest Ring of Gyges experiment ever conducted. Billions of participants across every culture and every continent entered the experiment over the course of three decades. There was no control group, no informed consent, no ethics board reviewing the protocol. The experiment simply happened, propelled by market forces and technological momentum and the genuine belief of its architects that the results would be liberating.

The results were for some people, in some ways, for some period of time. Political dissidents in authoritarian regimes used the internet's anonymity to organize resistance. Marginalized communities found each other across distances that would have kept them isolated in the physical world. Whistleblowers exposed corruption. Independent journalists reported stories that the mainstream press would not touch. The internet's capacity to amplify voices that the physical world had silenced was real. It was not an illusion.

But the experiment also generated data that Plato would have recognized. The data showed that anonymity plus audience produced predictable patterns of escalation. It showed that communities built on the absence of identity struggled to maintain trust at scale. It revealed that the gap between the person and the persona could be filled with liberation or with cruelty. That the architecture of the platform determined which tendency prevailed. Above all, the data showed that human nature did not improve when the body was removed from the conversation. Human nature simply expressed itself more freely, in all of its contradictory fullness.

The early internet was an age of innocence, and like all ages of innocence, it ended not with a single catastrophe but with a gradual accumulation of evidence. The evidence arrived in flame wars and troll campaigns, in harassment and deception, in the slow erosion of the norms that small communities had maintained through social pressure alone. But it arrived alongside genuine beauty and genuine connection and genuine acts of generosity that the physical world could not have produced. Both were real. Both were products of the same technology.

Unlike Plato's thought experiment, the internet generates data. Billions of interactions, recorded and archived, constitute the largest dataset in human history on the question that Glaucon posed in the Republic. What do people become when no one is watching? The answer, it turns out, is not a single thing. The answer is everything. The internet revealed that human beings are capable of extraordinary generosity when no one knows their name. They are capable of extraordinary cruelty under the same conditions.

Peter Steiner earned a total of $150 from his original submission to *The New Yorker*.[1] He later renegotiated the terms as the cartoon's fame grew. By the 2000s, it had appeared in more than a hundred publications. Steiner told interviewers that he found the cartoon's persistence somewhat bewildering. He had drawn it as a joke. He had not intended it as a philosophical statement. But the best prophecies are the ones that arrive disguised as jokes. They slip past the defenses of seriousness. They lodge in the memory because they are funny. And then, years later, you remember the punchline and realize it was never really a punchline at all. It was a diagnosis. On the internet, nobody knows you're a dog. Who you choose to be on the internet can change like the wind. Ultimately, it's up to the individual to decide.

References

[1] Peter Steiner, cartoon, *The New Yorker*, July 5, 1993, 61. Details of the cartoon's publication history, Steiner's fee, and his subsequent

reflections are drawn from Michael Cavna, "The Dog That Launched a Thousand Memes," *Washington Post*, October 31, 2013.

[2] John Perry Barlow, "A Declaration of the Independence of Cyberspace," February 8, 1996. Originally distributed via email and widely archived online. Full text available at the Electronic Frontier Foundation, eff.org.

[3] On the origins of ARPANET, see Katie Hafner and Matthew Lyon, *Where Wizards Stay Up Late: The Origins of the Internet* (New York: Simon & Schuster, 1996). The hacker ethic is codified in Steven Levy, *Hackers: Heroes of the Computer Revolution* (Garden City, NY: Anchor Press/Doubleday, 1984). On the September 1993 AOL gateway to Usenet and its cultural impact, see Dave Fischer, "Eternal September," alt.folklore.computers, January 26, 1994.

[4] Stewart Brand and Larry Brilliant co-founded the Whole Earth 'Lectronic Link (the WELL) in 1985. Howard Rheingold, *The Virtual Community: Homesteading on the Electronic Frontier* (Reading, MA: Addison-Wesley, 1993). The "You Own Your Own Words" principle and Rheingold's reflections on the WELL's internal tensions are drawn from this work and from his subsequent essays on the subject.

[5] United States v. Alkhabaz (also known as United States v. Baker), 104 F.3d 1492 (6th Cir. 1997). The case history, Baker's postings to alt.sex.stories, and Judge Avern Cohn's ruling are drawn from court records and contemporaneous reporting, including Peter H. Lewis, "An

Internet Author of Prose So Violent as to Unsettle the Most Hardened," *New York Times*, February 11, 1995.

[6] Mike Godwin, "Meme, Counter-Meme," *Wired*, October 1994. Godwin first formulated his law of online discourse in 1990 on Usenet. See also Mike Godwin, *Cyber Rights: Defending Free Speech in the Digital Age* (New York: Times Books, 1998).

Chapter Four - The Mask Effect

In 2004, a psychologist named John Suler sat down to describe something that millions of people had already felt but could not articulate. He was a professor at Rider University in New Jersey, and his research focused on the psychology of cyberspace. The internet was barely a decade old as a consumer phenomenon. Chat rooms and message boards had been thriving for years. People who would never raise their voice at a dinner party were typing obscenities at strangers. People who were shy in hallways were confessing their deepest fears to faceless avatars. Something was happening to human behavior when the screen became a barrier between one person and another. Suler wanted to name it.

He called it the Online Disinhibition Effect.[1] The term was clinical and precise, but the phenomenon it described was anything but orderly. Suler argued that the internet did not simply give people a new place to communicate. It gave them a new psychological environment, one that systematically dismantled the internal restraints that govern face-to-face interaction. He identified six distinct factors that combined to produce this disinhibition. Each one operated independently. Together, they formed a mechanism so powerful that it could transform the behavior of otherwise ordinary people within minutes of sitting down at a keyboard.

Suler's six factors were not abstract theories. They were descriptions of what was already happening across every corner of the early internet. They were also, whether Suler intended the comparison or not, a modern translation of the thought experiment that Glaucon proposed in Plato's Republic centuries earlier. Glaucon argued that invisibility would strip away the moral behavior of any person. Suler identified exactly how that stripping occurs in the digital world. Where Glaucon saw one ring, Suler saw six interlocking mechanisms. Each one deserves attention because each one is operating on you right now, every time you open a browser or unlock your phone.

The first factor is dissociative anonymity. "You don't know who I am." This is the ring itself in its purest form. When a person operates under a username or behind the shield of an anonymous account, the actions taken in that space feel separated from the person who takes them. The username becomes a character. The real self watches from a safe distance. A middle school teacher in Ohio can spend her evenings writing vicious political commentary under the handle "PatriotMom1776" and feel no contradiction between her classroom persona and her online one. The anonymous identity absorbs the behavior. The real identity stays clean.[1]

The second factor is invisibility. "You can't see me." This goes beyond anonymity because even people who use their real names online are still invisible in the physical sense. There is no eye contact. There is no body language to read. The person on the other end of the conversation exists only as text on a screen. Research has consistently shown that eye contact is one of the most powerful regulators of human social behavior.[7] When you can see the face of the person you are speaking to, mirror neurons fire and empathy engages automatically. Remove the face and you remove that automatic empathy response. The other person becomes an abstraction.

The third factor is asynchronicity. "I don't have to deal with your immediate reaction." In a face-to-face conversation, the consequences of your words arrive instantly. You see the wince, the tears, the anger. Online communication breaks this feedback loop. A person can type something cruel, press send, and close the laptop. The reaction happens somewhere else, to someone else, at some other time. By the time the target reads the message and responds, the sender may be eating dinner or watching television. The emotional weight of the exchange is distributed across time in a way that dilutes accountability.[1]

The fourth factor is solipsistic introjection. "I'm not talking to a real person." This is perhaps the most unsettling of Suler's six mechanisms because it operates below conscious awareness. When people communicate online, they often construct a mental model of the person they are talking to. That model is built from fragments. A username, a profile picture, a few lines of text. The mind fills in the rest based on its own expectations and projections. The result is that many online conversations are not actually between two people. They are between one person and a character that the other person has constructed inside their own head. Cruelty becomes easier when the target of that cruelty is not a real human being but a mental puppet that the aggressor is manipulating without realizing it.[1]

The fifth factor is dissociative imagination. "This is a game, not real life." Many people experience the online world as a separate realm with its own rules. What happens on the internet stays on the internet. The argument in the comments section is not a real argument. The insult hurled at a stranger on Twitter is not a real insult. This psychological partitioning allows people to behave in ways they would never tolerate in themselves if they believed the behavior counted. The screen becomes a stage. The person typing becomes a performer. The performance ends when the browser closes.[1]

The sixth factor is the minimization of authority. "No one is in charge here." In physical spaces, authority figures regulate behavior through their mere presence. Students behave differently when the teacher is in the room. Employees moderate their speech when the boss walks by. Online spaces strip away these authority cues. The CEO and the intern occupy the same comment thread with equal standing. The professor and the first-year student trade insults on equal footing. Without visible authority, the social hierarchy that normally constrains behavior collapses. People say things they would never say in any room where someone with power over them might be listening.[1]

These six factors do not require all six to be present for disinhibition to occur. Any combination will produce some degree of the effect. But when all six converge, the result is a psychological environment that Glaucon would have recognized immediately. It is the ring of Gyges, not as a mythological artifact but as an architectural feature of the internet itself. Every platform ever built contains some combination of these six elements. The variation is only in degree.

Thirty-five years before Suler published his framework, a young psychologist at New York University was already testing the core of this idea in a basement laboratory. Philip Zimbardo was not interested in the internet. The internet would not exist as a public network for another quarter century. Zimbardo was interested in something more fundamental. He wanted to know what happens to ordinary people when you strip away their individual identity, what emerges when a person stops being a specific someone and becomes part of an anonymous group. His tools were not screens or keyboards. They were something much simpler and much older. He used hoods.

The year was 1969. Zimbardo recruited female college students for what he described as a study on empathy and decision-making.[3] The women arrived at the laboratory expecting something routine. They were psychology students. They had participated in studies before. Nothing about the initial briefing would have suggested that this particular afternoon would produce data still cited more than half a century later. The participants were divided into two groups. The first group wore large white hoods that covered their faces and oversized lab coats that obscured their bodies. They sat in a dimly lit room where shadows pooled in the corners. No name tags were visible. The researcher addressed them only as a group, never as individuals. They could not see each other's expressions beneath the hoods. They were a collection of anonymous shapes in a dark room. The second group wore their normal clothes. They had large name tags pinned to their chests. The room was brightly lit. The researcher addressed them by name, making eye contact as she spoke to each one.

Both groups were given the same task. They were told that they would be evaluating another participant in a separate room. They could hear this person through an intercom. Their evaluation tool was an electric shock button. They were instructed to administer shocks as part of the "learning task" whenever they felt it was appropriate. The shocks were not real, but the participants believed they were. They could hear the reactions of the person in the other room, who was actually a confederate following a script.

The results were stark. Hooded women administered electric shocks that lasted nearly twice as long as those delivered by the women wearing name tags.[3] The difference was not subtle. It was not a marginal statistical blip that required sophisticated analysis to detect. The hooded group held the button down with a consistency that suggested something fundamental had shifted in their decision-making. The hoods had been on their faces for less than an hour. No one had threatened them. No one had offered them money or approval for cruelty. No one had told them to be harsh. The only variable was whether they believed they could be identified as individuals. When the answer was no, their willingness to inflict pain on a stranger nearly doubled.

Zimbardo's interpretation was direct. He called the phenomenon deindividuation.[3] When a person loses their sense of individual identity, the internal constraints that normally regulate their behavior weaken dramatically. The hooded women were not sadists. They were ordinary college students who had walked into a lab that morning with no intention of hurting anyone. The hoods did not change who they were. The hoods changed what they were willing to do. Two years later,

Zimbardo tested the same principle on a much larger and more terrifying scale.

In August of 1971, Zimbardo transformed the basement of Stanford University's psychology building into a mock prison.[4] He recruited twenty-four male college students who had been screened for psychological stability and randomly assigned them to play either guards or prisoners. The guards were given khaki uniforms and wooden batons. They were also given mirrored sunglasses that hid their eyes. This detail would prove to be one of the most significant design choices in the history of experimental psychology.

The experiment was supposed to last two weeks. It lasted six days.[4] On the first morning, the local police department assisted Zimbardo by conducting mock arrests of the students assigned to be prisoners. Officers showed up at their homes without warning, handcuffed them, read them their rights, and drove them to the campus in squad cars. The prisoners were stripped, deloused, given smocks with identification numbers, and placed in cells constructed from converted laboratory rooms. The guards received no specific training in how to maintain order. They were simply told to keep the prison running. Zimbardo wanted to see what would emerge naturally from the structure of the situation itself.

On the first day, the interactions were awkward and self-conscious. The guards and prisoners were still college students playing roles. Both groups laughed nervously. The guards tested their authority with small requests. The prisoners complied with mild grumbling. By the second day, something shifted. The guards began enforcing arbitrary rules with increasing aggression. They forced prisoners to do push-ups as punishment. They confiscated mattresses from cells. They restricted bathroom access. By the third day, the guards were waking prisoners in the middle of the night for "counts" that served no purpose except to demonstrate power. One prisoner had a breakdown so severe that he had to be released from the study. Two more followed within the next forty-eight hours.

The mirrored sunglasses mattered. They gave the guards the same psychological cover that the white hoods had given to the women in the 1969 experiment. The prisoners could not see the guards' eyes. The guards could not be identified through eye contact. This small barrier created the same dissociative effect that Suler would later describe as the combination of anonymity and invisibility. The guards were not being watched as individuals. They were performing a role behind a mask. The mask liberated them from the social expectations that would have governed their behavior in any other context.

Zimbardo himself became absorbed in the experiment. He later admitted that he lost perspective on the suffering of the participants.[2] He had started the study as a researcher. Within days, he was thinking of himself as the superintendent of a real prison. He worried about security and escapes. The psychological damage being inflicted on the young men in his care stopped registering. It took an outside observer, a graduate student named Christina Maslach who was not part of the research team, to walk into the basement and see what everyone inside had stopped seeing. She told Zimbardo that what was happening was wrong. She told him that he needed to end it. He terminated the experiment the next morning.

The Stanford Prison Experiment became one of the most famous studies in the history of psychology. It also became one of the most criticized.[2] Subsequent researchers pointed out the small sample size. They questioned whether the guards were responding to demand characteristics, meaning they were behaving the way they thought Zimbardo wanted them to behave. A 2002 BBC replication of the experiment produced different results, with the guards showing less aggression and the prisoners organizing collective resistance.[2] The replication debate continues to this day.

These criticisms are legitimate and they deserve serious engagement. But the directional finding has survived decades of scrutiny.[6] The specific details of the Stanford experiment may be contested. The broader principle that deindividuation increases antisocial behavior has been confirmed across dozens of subsequent studies using different methods and different populations. A 1998 meta-analysis by Tom Postmes and Russell Spears examined the full body of deindividuation research and concluded that the removal of individual identity consistently shifts behavior toward the extremes of whatever norm is present in the group.[6] If the group norm is cruelty, deindividuation amplifies cruelty. If the group norm is generosity, deindividuation amplifies generosity. The mask does not point in one direction. It amplifies whatever direction is already present.

This finding matters enormously for understanding what the internet does to human behavior. Zimbardo was essentially running Glaucon's thought experiment in a controlled laboratory setting. He gave ordinary people a version of the ring. The hoods and the sunglasses were crude analogs for the invisibility that Gyges enjoyed. The results confirmed that Glaucon was largely right. Remove accountability and behavior changes.

Remove individual identity and the restraints fall away. The ring works.

If anonymity were the entire explanation, then the solution would be simple. Remove anonymity. Require real names. Make everyone visible. Force every internet user to stand in the bright room with the name tag on their chest. The logic is intuitive and it has appealed to policymakers, platform designers, and concerned citizens for decades. If people behave badly because they are anonymous, then end the anonymity. The problem will solve itself.

South Korea tried exactly this on a national scale. The results were illuminating and deeply uncomfortable for anyone who believed that transparency alone could fix the problem.

In 2007, the South Korean government implemented what became known as the Real-Name Internet Policy.[5] The law required users to verify their real identities before posting comments on major websites with more than 100,000 daily visitors. The policy was a direct response to a wave of cyberbullying incidents that had received intense media coverage. Several high-profile cases involved celebrities who had been driven to suicide by anonymous online harassment. The public demanded action. The government responded with what seemed like a logical intervention. If people had to use their real names, they would behave better.

Researchers Daegon Cho and Soodong Kim studied the effects of this policy across multiple platforms over several years.[5] Their findings were released in 2012, and they delivered a result that surprised nearly everyone. Malicious comments dropped by approximately 0.9 percent. Not nine percent. Zero point nine percent. The real-name requirement had almost no measurable effect on the behavior it was designed to prevent.

The explanation was hiding in Suler's framework all along. Anonymity is only one of six factors that produce online disinhibition. The South Korean policy addressed the first factor, dissociative anonymity, by forcing people to attach their real names to their words. But it did nothing about the remaining five. The commenters still could not see the people they were attacking. They still did not have to face an immediate emotional reaction. Mental models of their targets still replaced engagement with real human beings. They still experienced the internet as a separate realm from physical life. Authority figures were still absent from the comment sections.

The ring of Gyges is not just a name. It is a psychological architecture. Removing one component leaves five others intact. The South Korean experiment proved that you cannot dismantle the ring by pulling out a single piece.

The policy was eventually repealed in 2012 after a massive data breach exposed the real identities of millions of users.[5] The government had collected an enormous database of verified names linked to online accounts. Hackers stole that database. A policy that was supposed to make people safer had instead created a new and more dangerous form of vulnerability. The real names that were supposed to enforce accountability became weapons in the hands of criminals who could now link anonymous posts to physical identities. In the end, the cure had been worse than the disease.

Facebook's experience tells a parallel story. Since its founding, Facebook has maintained a real-name policy that requires users to register under their legal names. The platform was designed from the beginning as an identity-verified social network. Users post under their own names alongside photos of their families. Their coworkers can see their comments. Their mothers can read their posts. If real names were sufficient to prevent harmful behavior, Facebook should have been a haven of civil discourse. The reality has been almost precisely the opposite. Facebook has been plagued by harassment, hate speech, coordinated bullying, and the spread of dangerous misinformation since its earliest days as a public platform. Internal research leaked in 2021 revealed that the company's own data scientists had documented the platform's role in amplifying divisive content. The real-name requirement did not prevent any of it because the other five factors of Suler's framework were still fully operational on every page of the site. Users could still avoid eye contact. They could still post and walk away. They still constructed mental models of their opponents rather than engaging with real people. The physical distance between commenter and target remained infinite. Comparison data between Facebook and anonymous platforms like 4chan or Reddit showed that while the nature of harassment differed across platforms, the overall prevalence remained stubbornly high regardless of whether real names were attached to the behavior.

This convergence of evidence from controlled experiments, national policy initiatives, and platform design points toward a conclusion that is both obvious and frequently ignored. Real names are not the answer. Visibility is not the answer. No single intervention can neutralize the disinhibition effect because the effect is produced by multiple independent mechanisms operating simultaneously.

In 2004, the same year that Suler published his academic framework, two webcomic artists named Mike Krahulik and Jerry Holkins published their own version of the same insight. They called it the Greater Internet Fuckwad Theory.[9] The formula was simple. A normal person, plus anonymity, plus an audience, equals a total fuckwad. The language was crude and deliberately provocative. The observation was precise.

The theory went viral almost immediately. It was shared across forums and blogs and became one of the earliest pieces of internet folk wisdom.[9] People recognized the truth in it because they had seen it happen. They had watched it happen to their friends. They had watched it happen to themselves. A reasonable person sits down at a keyboard. An audience gathers. Accountability vanishes. The reasonable person says something they would never say to anyone's face. Krahulik and Holkins had no training in psychology. They were cartoonists who drew a webcomic about video games. But they had identified the same core dynamic that Suler described in clinical terms and that Zimbardo had demonstrated in a laboratory. The ring turns ordinary people into something else. The audience gives them a reason to perform. The invisibility gives them permission.

What Krahulik and Holkins added to the framework, almost by accident, was the role of the audience. Suler's six factors described the conditions that remove restraint. The audience provides motivation. Cruelty performed in private is just cruelty. Cruelty performed before a crowd becomes a display of dominance. The anonymous commenter who tears apart a stranger's post is not doing it in silence. The commenter is doing it in front of hundreds or thousands of other users who might upvote, share, or applaud the attack. The ring removes the cost of cruelty.

The audience provides the reward.

The individual studies and case examples tell vivid stories about specific people in specific rooms making specific choices. The aggregate data tells a more sobering story about what is happening across entire populations.

The Pew Research Center has been tracking online harassment in the United States since 2014.[8] Their surveys paint a picture of a phenomenon that is not marginal or exceptional. It is pervasive. In their 2021 survey, 41 percent of American adults reported that they had personally experienced some form of online harassment.[8] This was not a narrow definition. It included name-calling, deliberate embarrassment, physical threats, sustained harassment campaigns, stalking, and sexual harassment.

Twenty-five percent of American adults reported experiencing what Pew classified as severe harassment.[8] This category included physical threats, stalking, sustained harassment over a period of time, and sexual harassment. One in four adults. In a country of roughly 260 million adults, that figure translates to approximately 65 million people who have experienced someone threatening their physical safety, following them across platforms, or subjecting them to sustained campaigns of abuse. These were not isolated incidents involving a handful of trolls. This was a mass behavioral phenomenon affecting tens of millions of people. The 2017 survey had shown similar numbers. The 2014 survey had shown similar numbers. The trend line was not improving. If anything, the problem was calcifying into a permanent feature of online life.

The data also revealed patterns that matched the predictions of Suler's framework with troubling precision. Harassment was not distributed equally across platforms or populations. Platforms with greater anonymity and weaker moderation produced higher rates of harmful behavior. Platforms with real-name requirements produced lower rates, but the reduction was modest rather than transformative. The gap between anonymous and identified platforms was real but not nearly as large as most people assume.[8]

The targeting of harassment followed demographic fault lines that predated the internet by centuries. Women were more likely than men to experience sexual harassment online. Young women between the ages of 18 and 29 reported the highest rates of any demographic group. People of color were more likely to experience harassment related to their race or ethnicity. LGBTQ individuals were more likely to be targeted for their sexual orientation or gender identity.[8] The ring did not create these prejudices. It amplified them. The biases existed before the internet. Misogyny existed before the first modem was plugged in. Racism existed before the first chat room opened. The internet provided a low-cost, low-risk environment in which those biases could be expressed without consequence. The mask gave old hatreds a new and frictionless delivery system.

Academic research has confirmed these patterns through more rigorous methods. In 1998, Tom Postmes and Russell Spears published a comprehensive meta-analysis of deindividuation research that examined studies spanning decades.[6] Their central finding was nuanced and important. Deindividuation did not simply make people more aggressive. It made people more responsive to situational norms. In contexts where the prevailing norm was aggression, deindividuation increased aggression. In contexts where the prevailing norm was cooperation, deindividuation increased cooperation. The mask did not have a single effect. It had an amplifying effect that depended entirely on what was already in the room.

In 2012, Noam Lapidot-Lefler and Azy Barak conducted an experiment that isolated the specific contributions of anonymity, invisibility, and lack of eye contact to toxic online behavior.[7] They designed an online interaction in which participants discussed controversial topics under varying conditions. Some participants were anonymous. Some were visible via webcam. Some could see each other's eyes. The researchers measured the frequency and intensity of hostile statements across all conditions.

The results were revealing. Anonymity alone produced a modest increase in hostile behavior. Invisibility alone produced a similar modest increase. But the absence of eye contact produced the largest single effect.[7] When participants could not see each other's eyes, hostile behavior increased most dramatically. This finding aligned with decades of research on the role of eye contact in social regulation. The eyes are the primary channel through which human beings read emotion in one another. Remove the eyes and you remove the strongest natural brake on cruelty.

The most powerful predictor of hostile behavior was the combination of all three factors.[7] When participants were anonymous, invisible, and unable to make eye contact, toxic behavior reached its peak. This finding mapped precisely onto Suler's framework. No single factor was sufficient. The combination was devastating. The pattern that emerges across all of this data is consistent and clear. Anonymity alone increases harmful behavior, but only modestly. Anonymity combined with invisibility increases it further. Anonymity combined with invisibility, audience, and the absence of eye contact produces the strongest effect. The ring of Gyges is not a single mechanism. It is a system of reinforcing mechanisms, each one lowering a different barrier, each one removing a different form of friction that would otherwise keep behavior in check.

All of this evidence circles back to a question that has been lurking beneath every experiment, every data set, and every case study in this chapter. It is the question that has shadowed this book from its opening pages. It is the question that Glaucon posed in the Republic and that we have been wrestling with ever since. Does the mask change people or does it reveal them?

The question sounds simple. The implications are enormous. If the mask changes people, then Glaucon was wrong. Human beings are naturally good, or at least naturally restrained, and the removal of accountability corrupts them. The anonymous commenter who writes hateful things is not expressing their true self. They are expressing a distorted version of themselves that the technology has produced. The cruelty is an artifact. The goodness is the baseline.

If the mask reveals people, then Glaucon was right. Human beings are naturally selfish and aggressive. They behave well only because they fear the consequences of behaving badly. The anonymous commenter is the real person. The polite colleague at the office is the performance. The cruelty is the baseline. The goodness is the artifact.

The evidence presented in this chapter does not support either of these clean narratives. It supports something messier, more nuanced, and ultimately more disturbing.

The hooded women in Zimbardo's 1969 experiment were not naturally cruel. They were not performing cruelty because the experiment demanded it. Other participants in the same experiment, also wearing hoods, did not increase their shock durations.[3] The effect was real, but it was not universal. Some people resisted the pull of the mask. Others did not. The same variation appeared in the Stanford Prison Experiment. Some guards became abusive almost immediately. Others remained relatively restrained throughout the study.[4] The mask did not produce a uniform result. It produced a range of results that depended on something inside each individual.

The Postmes and Spears meta-analysis provides the key to understanding this variation.[6] Deindividuation does not push everyone in the same direction. It amplifies whatever direction is already present. If the situational norms point toward cruelty, the mask amplifies cruelty. If the norms point toward kindness, the mask amplifies kindness. The mask is not a moral force. It is a moral amplifier.

This means that the answer to the question of whether the mask changes people or reveals them is neither. Every person carries within them a range of possible behaviors. Some of those behaviors are generous and compassionate. Some are petty and cruel. Some are creative and others are destructive. In the course of a normal day, the context determines which possibilities get expressed. The presence of other people, the threat of consequences, the norms of the environment, the authority structures in place... all of these factors shape which version of the self emerges at any given moment. The mask removes many of these contextual constraints simultaneously. It strips away the presence of other people through invisibility. It eliminates the threat of consequences through anonymity. Authority dissolves through the flattened hierarchy of the internet. The empathy circuit breaks through the absence of eye contact. What remains after all of these removals is not the "true self" in any meaningful sense. What remains is whichever impulse happens to be strongest at that particular moment in that particular context.

This is the third possibility that neither Glaucon nor Socrates considered. People are not good. Nor are they evil. They are multiple things simultaneously, with the balance between those things constantly shifting based on circumstances. The mask does not reveal a fixed inner nature. It creates conditions in which the worst impulses have the least friction and the best impulses have the least support. Suler himself recognized this duality. He noted that the online disinhibition effect cuts in two directions.[1] There is toxic disinhibition, which produces cruelty and aggression. There is also benign disinhibition, which produces unusual honesty, emotional openness, and acts of generosity that the person might never attempt face to face. The same anonymity that allows a stranger to send a death threat also allows a teenager to confess their sexuality in an online support group. The same invisibility that enables a bully also enables a grieving parent to seek comfort without shame. The ring gives power to whatever impulse reaches it first.

Consider what this means in practical terms. The same person who volunteers at a food bank on Saturday morning might write a vicious comment on a news article on Saturday evening. The same person who is patient with their children at breakfast might send a threatening message to a stranger before lunch. These are not contradictions if you accept the multiplicity model. They are different selections from the same range of possibilities, triggered by different contextual conditions. The food bank provides face-to-face contact, social accountability, authority structures, and immediate emotional feedback. A comment section provides none of these things. The ring does not transform Dr. Jekyll into Mr. Hyde. It creates conditions in which Mr. Hyde faces no resistance.

This understanding is worse than Glaucon's binary in at least one important respect. Glaucon's version of the argument allowed for a clean prediction. Give anyone the ring and they will become unjust. That prediction is terrifying, but it is also simplistic. It allows you to build defenses against a known threat. The multiplicity model allows for no such certainty. It means the answer to "Would you use the ring for evil?" is not yes or no. The answer is "It depends on the day." It depends on whether you slept well. It depends on whether you are angry about something that happened at work. The answer shifts if the person you encounter online reminds you of someone you dislike. It depends on whether you are lonely or bored or afraid. A thousand contextual variables that shift from hour to hour, most of which you are not even conscious of when you sit down at your keyboard.

The ring does not create evil. It removes the friction that keeps evil contained. The amount of friction required varies from person to person and from moment to moment within the same person. On a good day, the friction holds. On a bad day, it may not. This is not a comfortable conclusion. It offers no clean villains and no clean heroes.

Suler's six factors are not a description of what the internet does to bad people. They are a description of what the internet does to all people.[1] Zimbardo's experiments are not a warning about a rare form of human pathology. They are a demonstration of a universal human vulnerability.[2] The Pew data showing that 41 percent of American adults have experienced online harassment is not a statistic about a troubled minority of internet users. It is a statistic about the conditions under which ordinary behavior becomes extraordinary.[8]

The ring is always on our finger now. We wear it every time we open a browser or type a response to a stranger. We wear it every time we scroll through a feed and feel the distance between ourselves and the faces we cannot see. Billions of people put on the ring every morning when they check their phones. They wear it through the day as they comment, react, share, and argue with people they have never met and will never meet. They take it off only when they close the screen and return to the physical world where faces are visible and consequences are immediate. Some people never take it off at all. The question is no longer whether the ring exists. The question is what we become on the days when its power finds us at our weakest.

And the honest answer... the one that the data compels and that Glaucon would have appreciated for its unflinching clarity... is that we do not know. We do not know what we will become until the moment arrives. The woman in the hood did not know she would hold the shock button longer than she ever would have without it. The guard in the mirrored sunglasses did not know he would wake a stranger at three in the morning to assert dominance over a college student in a smock. The commenter in the thread did not know they would type something they would be ashamed to read aloud to their own family. None of them knew until the mask was on and the moment presented itself. The mask does not determine the outcome. It only determines the stakes. Everything else depends on which version of ourselves shows up to wear it.

References

[1] Suler, John. "The Online Disinhibition Effect." *CyberPsychology & Behavior* 7, no. 3 (2004): 321-326.

[2] Zimbardo, Philip. *The Lucifer Effect: Understanding How Good People Turn Evil*. Random House, 2007.

[3] Zimbardo, Philip. "The Human Choice: Individuation, Reason, and Order vs. Deindividuation, Impulse, and Chaos." In *Nebraska Symposium on Motivation*, 1969.

[4] Haney, Craig, Curtis Banks, and Philip Zimbardo. "Interpersonal Dynamics in a Simulated Prison." *International Journal of Criminology and Penology* 1 (1973): 69-97.

[5] Cho, Daegon, and Soodong Kim. "Empirical Analysis of Online Anonymity and User Behaviors." *Hawaii International Conference on System Sciences*, 2012.

[6] Postmes, Tom, and Russell Spears. "Deindividuation and Antinormative Behavior: A Meta-Analysis." *Psychological Bulletin* 123, no. 3 (1998): 238-259.

[7] Lapidot-Lefler, Noam, and Azy Barak. "Effects of Anonymity, Invisibility, and Lack of Eye-Contact on Toxic Online Disinhibition." *Computers in Human Behavior* 28, no. 2 (2012): 434-443.

[8] Pew Research Center. "The State of Online Harassment." January 2021.

[9] Krahulik, Mike, and Jerry Holkins. "Green Blackboards (And Other Anomalies)." *Penny Arcade*, March 19, 2004.

Chapter Five - Hunting in Packs

This is the hardest chapter in this book to read. It contains stories of real people who were harassed, threatened, stalked, and destroyed by anonymous strangers on the internet. Some of those people are dead. The facts of their lives and deaths deserve to be told plainly, without embellishment. The events described here are sensational enough on their own. They do not need a writer's help to shock you.

Glaucon imagined one shepherd with one ring. He asked what a single person would do if freed from all consequence. The question was disturbing enough in that form. But Glaucon never imagined what would happen if millions of people put on the ring at the same time.

On August 16, 2014, a twenty-four-year-old computer programmer named Eron Gjoni published a 9,425-word blog post about his ex-girlfriend. The post was titled "The Zoe Post." It detailed his relationship with Zoe Quinn, an independent game developer who had created a small interactive fiction game called *Depression Quest*. The game was free to play. It explored the interior experience of living with clinical depression through a series of text-based choices. The game was not the kind that typically generates controversy. Most people never hear about it at all.

Gjoni's post accused Quinn of infidelity, naming several men in the gaming industry she had allegedly slept with. One of those men was a journalist who had written about her work. The post was written in a tone of wounded sincerity. It included screenshots of private text conversations. It presented itself as an act of transparency, a scorned boyfriend airing legitimate grievances for the public record. What it actually accomplished was the ignition of one of the most sustained coordinated harassment campaigns the internet had ever produced.[1]

Within hours, anonymous users on 4chan's v board began organizing. The v board was a video game discussion forum that required no registration and displayed no usernames. Every poster appeared as "Anonymous." The users seized on the allegation that Quinn had slept with a journalist as evidence that the gaming press was corrupt. The rallying cry became "ethics in games journalism." This phrase would serve as the movement's public face for years. Beneath that face lay something far less principled.[10]

The anonymous boards moved fast. Users on 4chan created threads dedicated to compiling Quinn's personal information. They found her home address. Her phone number surfaced next. Then came the names and addresses of her family members. All of this was published openly, an act known as "doxxing." The threads served as coordination centers where users discussed strategy, assigned targets, and celebrated successful attacks. Then the direct threats began.[1]

Quinn received death threats so specific that they included descriptions of how she would be killed. Rape threats followed, detailing the acts in graphic language. Other messages named her family members alongside promises to harm them. The volume was staggering. Hundreds of messages arrived per hour across every platform where she maintained a presence. Her email, her Twitter, her Tumblr, her personal website. All of them became conduits for abuse. She was driven from her home. She stayed with friends, then moved, then moved again. She later described the experience in her book *Crash Override*, writing that the harassment restructured her entire life around the need for physical safety.[1]

The coordination was deliberate and documented. Researchers who later analyzed the campaign found evidence of organized efforts on 4chan and the newer platform 8chan, where users planned specific actions and assigned targets in threads that functioned like operational briefings.[10] They created fake social media accounts to amplify their messages. Botnets flooded Quinn's mentions on Twitter. False reports were filed to get her accounts suspended. Participants contacted her employers and professional contacts with fabricated accusations. The mechanics were sophisticated, but the anonymity of the participants made each individual contribution appear trivial. Any single threatening message could be dismissed as the work of one disturbed person. But there was never just one message. There were thousands, arriving every day for months, then years.

Anita Sarkeesian was the next target. Sarkeesian was a media critic who had launched a Kickstarter-funded YouTube series called "Tropes vs. Women in Video Games." The series examined common patterns of female representation in games, such as the "damsel in distress" trope and the use of female characters as background decoration. It was academic in tone and moderate in its conclusions. Sarkeesian cited specific games and showed specific clips. She made arguments that would not have been controversial in a university media studies course. The response from anonymous corners of the internet was not moderate.

Sarkeesian had already been receiving threats since her Kickstarter campaign went viral in 2012. The Gamergate movement supercharged those threats into something unprecedented in scale. Someone created a browser game called "Beat Up Anita Sarkeesian" where players could click on her face until it was bloody and bruised. Her Wikipedia page was repeatedly vandalized with pornographic images. She received bomb threats at multiple speaking engagements, including one at Utah State University in October 2014 that promised "the deadliest school shooting in American history" if her lecture proceeded. The threat named Sarkeesian specifically. It promised a massacre of "feminists" in attendance.[2]

Utah State University officials consulted with local and federal law enforcement. Under Utah law, attendees with valid concealed carry permits could not be prevented from bringing firearms into the venue.

Sarkeesian made the decision to cancel the event. She testified before the United Nations Broadband Commission the following year, describing in measured language what her daily life had become. She told the assembled delegates that she had been driven from her home, that she had restricted her travel, that she had hired private security for public appearances, that she screened every piece of mail for threats before opening it. The cost of expressing opinions about video games while female had been calculated in death threats, bomb scares, and a constant undercurrent of fear.[2]

Brianna Wu received the same treatment on the same timeline. Wu was a game developer and head of development at Giant Spacekat, an independent studio. She had mocked a Gamergate meme on Twitter. The response arrived within hours. Anonymous users posted her home address along with explicit threats describing how they would rape and murder her. The threats were specific enough to include the layout of her neighborhood and a description of her house. Wu and her husband fled their home that night, packing bags in the dark and driving to a location they told no one about.[1]

The pattern across all three cases was identical in its structure and terrifying in its efficiency. A woman expressed an opinion or created something in the gaming space. Anonymous users identified her as a target. Coordination happened on anonymous platforms where no one used their real name. Personal information was compiled and published. Threats were generated at industrial scale. The target's life was disrupted, her safety was compromised, and her ability to work in her chosen field was severely damaged.

The shield that protected every participant was elegant in its simplicity. Every individual could claim plausible deniability. The person who found Quinn's address was "just sharing public information." Another who sent one angry message was "just expressing an opinion." The person who created a mocking meme was "just making a joke." Someone who used the hashtag was "just asking questions about ethics in journalism." Each action was minor in isolation. Each participant could look at their individual contribution and honestly say it was not, by itself, a serious harm. The aggregate effect was a sustained assault that lasted years and permanently altered the lives of its targets.

No single person bore responsibility for what happened. Every person was responsible for what happened. The ring had been distributed to thousands of hands, and the weight of the weapon they collectively wielded was too diffuse for any single hand to feel.

The most extreme manifestation of anonymous digital violence has a specific name. It is called swatting. The mechanics are straightforward. A person makes a phone call to an emergency dispatcher. They claim to be at a specific address where a violent crime is in progress. They describe an active shooter, a hostage situation, or a bomb threat. Police respond with maximum force, because that is what their training requires. A SWAT team arrives at the address with military-grade equipment, battering rams, flash grenades, and assault rifles. The person who lives at the address opens their door to find a dozen heavily armed officers pointing weapons at their face. They have no idea why this is happening. They have done nothing wrong. The caller is anonymous. The victim is not.

On the evening of December 28, 2017, a twenty-five-year-old man named Tyler Barriss sat in his apartment in Los Angeles, California, and picked up his phone. He called the Wichita, Kansas, emergency dispatch line. Speaking in a voice that conveyed convincing distress, he identified himself as a man at 1033 West McCormick Street who had just shot his father in the head. He said he was holding his mother and younger brother at gunpoint. He claimed to have poured gasoline throughout the house and said he was thinking about lighting the whole place on fire.[3]

The Wichita Police Department dispatched officers immediately. Multiple units converged on the address. Officers established a perimeter around the house with weapons drawn. At approximately 6:20 p.m. Central Time, Andrew Finch opened his front door to investigate the flashing lights and commotion outside. He was twenty-eight years old, a father of two young children who lived in the house with his mother and his niece. He had no connection whatsoever to the dispute that led to the phone call. He had never played Call of Duty. The name Tyler Barriss meant nothing to him.[3]

Officers shouted commands. Accounts differ on what happened in the next few seconds. An officer on the perimeter fired a single shot. Andrew Finch collapsed on his front porch. He was transported to a local hospital, where he was pronounced dead. His mother, who had come to the door behind him, watched him fall.[3]

The chain of events that killed Andrew Finch began with a dispute over a $1.50 wager in a Call of Duty online match. Two players named Shane Gaskill and Casey Viner had gotten into an argument during a game. The argument escalated from trash talk to personal threats. Viner contacted Tyler Barriss through an intermediary, because Barriss had earned a reputation in gaming circles as someone willing to swat people for a fee or simply for the thrill of doing it. Viner provided what he believed was Gaskill's home address. The address was wrong. It was Andrew Finch's address, a previous residence that Gaskill had since moved away from.[3]

Barriss had swatted dozens of people before that night. He had called in false reports to police departments across the country. His targets included private individuals, schools, television stations, a shopping mall, and a Federal Communications Commission meeting. He operated through layers of anonymity, using spoofed phone numbers that displayed false caller IDs and untraceable messaging accounts. Each call was a practiced performance. His voice mimicked distress convincingly enough to bypass dispatcher skepticism, and he knew the specific phrases and scenarios that would trigger the most aggressive armed response from law enforcement.[3]

The chain that killed Finch ran like this... an anonymous gaming dispute led to an anonymous threat, which led to an anonymous phone call, which triggered an armed police response, which produced a real bullet that entered a real body on a real front porch in Wichita, Kansas. Every link in that chain was insulated by anonymity except the final one. Andrew Finch was the only person in the entire sequence who was fully visible to the world. He was also the only person who died.

Barriss was eventually identified through digital forensics and arrested. He pleaded guilty to 51 federal charges and was sentenced to twenty years in prison. Casey Viner received a fifteen-month sentence. Shane Gaskill received deferred adjudication. But the Finch case was the exception in that anyone was caught at all. The FBI estimates that hundreds of swatting incidents occur in the United States every year.[3] The vast majority are never prosecuted because the callers use voice-over-IP services, disposable phone numbers, and anonymous messaging platforms that make identification extraordinarily difficult.

The targets of swatting include private individuals, members of Congress, judges, celebrities, and live-streaming gamers whose home addresses leak through data breaches or careless social media posts. In one particularly disturbing subset of cases, swatting has been used to interrupt live video streams. The audience watches in real time as armed police storm into the home of a person broadcasting from their bedroom or living room. The streamer's face registers confusion, then terror. Police shout commands. The chat window fills with messages. Some viewers express horror. Others express delight. They type laughing emojis and exclamation points while watching a human being held at gunpoint. The anonymous caller and the anonymous viewers together produce a spectacle of violence that treats a real person's terror as entertainment.

On April 15, 2013, two pressure cooker bombs detonated near the finish line of the Boston Marathon. Three people were killed and more than 260 were injured. The FBI released surveillance photographs of two unidentified suspects and asked the public for help identifying them. Thousands of anonymous Reddit users decided they could do better than the FBI.

A subreddit called r/findbostonbombers appeared within hours of the attack. It grew to tens of thousands of subscribers in days. Users began combing through thousands of photographs taken by spectators along the marathon route. They circled faces in red. Arrows pointed to suspicious-looking backpacks. Users compared jawlines and ear shapes and clothing brands. They built elaborate theories connecting specific individuals to the blast sites, complete with annotated maps and timestamp analyses. The effort was fueled by a collective conviction that the wisdom of the crowd would solve the crime faster than traditional law enforcement.[4]

The threads moved at extraordinary speed. Theories were proposed, upvoted, elaborated upon, and shared across platforms in a matter of hours. The normal checks that slow down a criminal investigation, the requirement for evidence, the presumption of innocence, the careful verification of witness statements, had no equivalent in the anonymous forum. There was only momentum.

On April 18, multiple Reddit threads converged on a name. Users identified a Brown University student named Sunil Tripathi as one of the bombing suspects. Tripathi had been reported missing by his family on March 16, a full month before the bombings. Users noted a perceived resemblance between Tripathi's missing person photograph and the FBI's grainy surveillance images. The identification spread from Reddit threads to Twitter feeds to mainstream news outlets within hours. Some reporters cited "social media" as their source without mentioning that the social media in question was an anonymous forum with no editorial standards.[4]

The Tripathi family's phone began ringing. The calls were not expressions of concern about their missing son. Anonymous callers threatened the family with violence. They received messages describing what should be done to them for raising a terrorist. Their voicemail filled with rage from strangers who had never met Sunil, who had no evidence connecting him to the bombing beyond a blurry photograph and the collective certainty of an anonymous crowd that had spent three days convincing itself it had cracked the case.[4]

Sunil Tripathi had nothing to do with the Boston Marathon bombing. The actual bombers were Tamerlan and Dzhokhar Tsarnaev, identified through law enforcement investigation and a dramatic manhunt that ended on April 19. Tripathi's body was found in Providence's India Point Park on April 25, 2013. The medical examiner determined he had died by suicide. The estimated date of death placed it before the bombing had even occurred. The Tripathi family had been mourning a lost son while simultaneously being accused of raising a murderer.[4] Erik Martin, Reddit's general manager at the time, issued a public apology. "We hoped that the crowdsourced investigation would not interfere with the official investigation," he wrote, "and we are very sorry for the pain caused to the Tripathi family."[4] The apology was notable for its understatement. An innocent family had been terrorized by thousands of anonymous strangers who were absolutely certain they had solved a crime. The crowd was wrong. The damage was irreversible.

The Boston Marathon misidentification was not an isolated failure of judgment. The pattern repeats wherever anonymous crowds appoint themselves as investigators and executioners. After every mass shooting, every terrorist attack, and every high-profile crime, the same sequence unfolds with depressing predictability. Anonymous users identify suspects based on fragmentary evidence. The identification goes viral. The named person and their family receive threats. Occasionally the identification turns out to be correct. Far more often it does not. The consequences of being wrong fall entirely on the accused and their families, while the accusers remain invisible and unaccountable.

The mechanism is simple enough to describe in a single sentence. Individual anonymity removes the fear of being wrong, collective conviction replaces the need for evidence, and the speed of digital sharing outpaces any possible correction. By the time the truth emerges, the damage has already been done. The innocent person has already been threatened. The family has already been terrorized. The retraction or correction never travels as fast or as far as the original accusation. The truth arrives on foot. The lie has been around the world twice by then.

No single Reddit user who identified Sunil Tripathi bore full responsibility for what happened to his family. Each one merely shared an observation or upvoted a theory or added a piece of speculative analysis. The collective effect was the terrorization of a grieving family and the desecration of a dead young man's memory. The ring made each participant invisible to consequence. The harm it produced was devastatingly visible to everyone except those who caused it.

In October 2012, a fifteen-year-old Canadian girl named Amanda Todd posted a nine-minute video to YouTube. The video was titled "My story: Struggling, bullying, suicide, self harm." In it, Todd held up a series of hand-written note cards that told her story in black marker on white paper. She never spoke a single word during the video. She let the cards speak for her. The camera showed her face, her hands, and the cards. That was all. The story the cards told began when Todd was in seventh grade. She had been using video chat to meet new people online. An anonymous stranger complimented her and convinced her to briefly expose herself on camera. The stranger captured the image. Then the stranger began to use it as a weapon.[5]

The anonymous predator tracked Todd across the internet with relentless persistence. Whenever she moved to a new school or created new social media accounts, the predator found her. He sent the captured image to her classmates, her teachers, and her parents' friends. He created a Facebook page using the image as a profile picture and sent it to everyone at her new school. The page arrived before Todd did, poisoning the social landscape of each fresh start before she had a chance to make it.[5]

Todd changed schools. The predator followed. She changed schools again. The predator followed again. At each new school, students who had never met her already knew her as the girl from the photograph. The bullying at school merged with the harassment online until there was no boundary between them. Classmates mocked her in hallways and on social media simultaneously. She was beaten up by a group of students from her school while other students filmed the attack on their phones and shared it online, where strangers joined in the mockery.

Todd attempted suicide by drinking bleach. She survived. The response from anonymous commenters online was not sympathy. They told her she should have used a different brand. They told her to try again and succeed this time.[5]

On October 10, 2012, roughly one month after posting her YouTube video, Amanda Todd died by suicide at her home in Port Coquitlam, British Columbia. She was fifteen years old. The anonymous predator who had initiated the cycle of abuse was eventually identified as Aydin Coban, a thirty-five-year-old Dutch national. Coban was arrested in the

Netherlands in 2014 and extradited to Canada, where he stood trial in British Columbia Supreme Court. He was convicted in 2022 on charges of extortion, criminal harassment, and possession and distribution of child sexual abuse material. The investigation revealed that Todd was far from his only victim. Coban had targeted dozens of young people across multiple countries, using the same pattern of anonymous contact, image capture, and relentless blackmail.[5]

Two years before Todd's death, a strikingly similar sequence unfolded at Rutgers University in New Jersey. Tyler Clementi was an eighteen-year-old freshman and a talented violinist. In September 2010, Clementi's roommate, Dharun Ravi, secretly used a webcam to film Clementi during an intimate encounter with another man in their shared dorm room. Ravi shared what he had seen with other students through text messages. He posted about it on Twitter. He set up the camera a second time and invited friends to watch a planned second encounter online.

Clementi discovered what his roommate had done. He checked Ravi's Twitter feed and read the messages for himself. On September 22, 2010, Clementi posted a final update to his Facebook page. "Jumping off the gw bridge sorry." He drove to the George Washington Bridge, climbed over the railing, and jumped into the Hudson River. His body was recovered six days later. He was eighteen years old.

These cases represent the extreme end of a spectrum that extends into millions of less visible lives. The Cyber Civil Rights Initiative has documented the scale of what it calls "non-consensual pornography," the distribution of intimate images without the subject's consent.[9] Their research indicates that approximately one in eight social media users has been a target of this practice. The vast majority of perpetrators are never identified because they operate behind layers of anonymity. The vast majority of victims are women and girls.

The phenomenon extends beyond image-based abuse into the broader category of sextortion, where anonymous predators use captured images or fabricated content to extort money or further sexual material from their targets. The FBI reported a dramatic increase in sextortion cases involving minors in recent years, with the perpetrators overwhelmingly operating from behind anonymous accounts.[9]

The common thread across every case in this chapter is a mechanism so simple it can be stated in a single breath. A person becomes visible as a target. Anonymous participants pile on. The cumulative weight of their individual actions destroys a life. The distance between each person's contribution and the final outcome is large enough that most participants never feel the weight of what they helped create.

The individual stories are devastating in their particulars. The aggregate numbers are devastating in their scale. As the Pew data discussed in the previous chapter showed, 41 percent of American adults reported experiencing some form of online harassment, with one in four experiencing severe forms including physical threats and stalking. In a country of roughly 260 million adults, the burden fell disproportionately on women, people of color, and LGBTQ+ individuals.[6]

The Anti-Defamation League's annual survey of online hate and harassment painted an even grimmer picture. Their 2023 report found that 52 percent of American adults had experienced some form of online harassment, a number that had been climbing steadily since they began tracking it. Among those harassed, 33 percent reported that the harassment was based on a core aspect of their identity. Their race, ethnicity, religion, gender, sexual orientation, or disability. The trend line across their annual reports pointed in one direction, and that direction was up.[7]

Seventy-five percent of harassment incidents went unreported to any authority or platform.[6] The reasons victims gave for not reporting followed a dispiriting pattern. They believed nothing would be done. Many feared that reporting would provoke retaliation from their harassers. Others did not know how to file a report or whom to file it with. Still others did not think the platform would care enough to act. These beliefs were, by and large, correct.

The pattern was not unique to the United States. Researchers in the United Kingdom documented similar rates of online abuse and similar failures of institutional response. South Korea experienced a crisis of cyber harassment so severe that it prompted sweeping national legislation. India saw cases of coordinated online harassment spilling into the physical world, with false rumors spread through WhatsApp groups triggering mob violence and lynchings in rural communities. Brazil documented targeted online campaigns against journalists and political figures that employed tactics indistinguishable from those of Gamergate, scaled up to a national level.[8]

The economics of online harassment illuminated a fundamental asymmetry that sat at the heart of the problem. For victims, the costs were enormous. The Cyber Civil Rights Initiative found that victims of non-consensual pornography suffered career disruption at high rates, with many reporting job loss or sustained difficulty finding employment.[9] Victims reported spending thousands of dollars on therapy. Some relocated to different cities to escape harassment that followed them into their offline lives. Some changed their legal names. The total financial burden of being targeted could reach into tens of thousands of dollars when accounting for legal fees, security measures, lost income, and mental health treatment.

For perpetrators, the cost was almost always zero. Sending an anonymous threat required no expenditure. Publishing someone's private address cost nothing. Creating a fake account to harass a stranger took seconds and demanded no investment beyond time. The fundamental imbalance was built into the architecture of anonymous digital communication. Causing harm was free, while enduring harm was financially and psychologically ruinous.

Danielle Keats Citron, a law professor who has studied online harassment for more than a decade, described this asymmetry in her influential book *Hate Crimes in Cyberspace*. She argued that the internet had created a new category of civil rights violation, one in which individuals were effectively driven from public participation by anonymous mobs.[8] The targets were disproportionately women and members of minority groups. The practical effect was a chilling of speech far more severe than any government censorship could achieve. People who knew they would be targeted for speaking simply stopped speaking. They deleted their social media accounts. Some withdrew from public forums. Others abandoned professional opportunities that required a public presence. Self-censorship became a survival strategy.

The invisible hand of the anonymous crowd was quietly reshaping who could participate in public life and who could not. The ring was not merely empowering individuals to behave badly when no one was watching. It was systematically removing certain categories of people from the public square.

When Glaucon proposed his thought experiment in the *Republic,* his point was that any person, given perfect immunity from consequence, would choose injustice over justice. The just man and the unjust man would behave identically once the risk of punishment disappeared. Glaucon's argument was provocative enough when applied to one person making one choice. It becomes something entirely different when you multiply it by a million.

Consider the arithmetic of anonymous harm. One anonymous person sends one threatening message to a stranger. The act is cruel. The harm is real but contained. The target reads one ugly message, feels one jolt of fear, and might reasonably dismiss it as the ravings of a single disturbed individual. Life goes on. The threat fades into the background noise of a difficult world.

Now multiply it. A thousand anonymous people send a thousand messages to the same target in a single day. The target wakes up to find their inbox overflowing. The messages come from different accounts across different platforms, written in different voices from different angles. Some are crude and barely literate. Some are articulate and chillingly specific. Some reference details about the target's daily routine that suggest sustained observation and careful planning. The target cannot dismiss a thousand messages as the ravings of one person. A thousand messages feel like a consensus. They feel like the world itself turning against you. They create the visceral sensation of being hunted by something vast and faceless and inescapable. This is the emergent property of collective anonymity. It creates harms that no individual participant feels responsible for, harms that no individual participant could have produced alone. The whole is catastrophically greater than the sum of its parts.

Each participant in the Gamergate campaign could honestly say "I just posted one comment." Each Reddit user who identified Sunil Tripathi could honestly say "I just shared my opinion." Each person who sent Amanda Todd a hateful message after her bleach attempt could honestly say "I just wrote one thing." Every single one of those statements was technically true. Every single one was also a component of a weapon that destroyed lives, disrupted careers, terrorized families, and in several documented cases ended in death.

Glaucon's shepherd acted alone. He made a conscious choice to do evil, and the evil he produced was proportional to his individual capacity for harm. The digital ring operates under no such limitations. When anonymity is distributed to millions of people simultaneously, the resulting capacity for harm scales beyond anything a single actor could achieve. No individual decides to destroy a life. The destruction emerges organically from the accumulated weight of millions of small choices, each one individually trivial, each one invisible to the person making it.

This is what Glaucon could not have foreseen. He understood that the ring's power lay in freeing the individual from external judgment. That insight was correct. But the far more dangerous power of the ring, the power that transforms it from a philosophical curiosity into an existential threat, is what happens when invisible people discover each other.

A single invisible shepherd is frightening in the way that a lone predator is frightening. A swarm of invisible shepherds, moving in loose coordination toward a common target, produces a category of harm that Glaucon's philosophy was not equipped to describe. The ancient world knew mobs. It knew riots and collective violence and the fury of crowds. But ancient mobs were visible to the world around them. Their participants could be identified afterward. They took physical risks when they gathered in the streets. They could be dispersed by authority. They could be held to account after the fact, however imperfectly.

The digital mob carries none of those costs and faces none of those risks. Its members are invisible. They gather without leaving their bedrooms. They act without endangering their own safety. And they disperse without trace. When the swarm moves on, the target is left standing alone in the wreckage, surrounded by damage that everyone caused and no one will claim.

The implications extend far beyond individual cases. When collective anonymous harassment becomes sufficiently common and sufficiently visible, it reshapes entire domains of public life through the power of example. Women leave social media at higher rates than men.[6] Female journalists self-censor to avoid being targeted. Game developers abandon projects rather than endure the gauntlet of anonymous abuse. Activists go silent. Politicians quietly modify their positions to avoid attracting the swarm's attention. The invisible crowd does not merely harm the specific individuals it targets. It alters the calculations and behavior of everyone who witnesses what the swarm can do.

Glaucon asked whether any individual person would remain just if freed from the consequences of injustice. The answer he proposed was no. The question this chapter forces us to confront is darker still. When millions of people are freed from consequence simultaneously, do they merely behave unjustly as individuals? Or do they produce a form of collective injustice that exceeds anything a single person could accomplish alone, a harm that is greater than the sum of its parts, a violence without a perpetrator?

The evidence assembled in the preceding pages suggests the latter. The ring does not simply empower isolated individuals to cause harm. It enables the formation of swarms. It creates a new category of collective violence that is simultaneously everyone's fault and no one's responsibility, a destruction that emerges from the aggregate the way weather emerges from the atmosphere. Andrew Finch is dead because an anonymous gamer made an anonymous phone call that ended in tragedy in Kansas. Amanda Todd is dead because an anonymous predator weaponized her image across anonymous platforms until the accumulated weight of collective cruelty became more than a fifteen-year-old girl could survive. The Tripathi family was terrorized because thousands of anonymous investigators on Reddit achieved absolute certainty about something that was absolutely false.

In each of these cases, the ring did not create a villain. It created a weather system. The harm arrived like a storm, generated by the accumulated energy of thousands of small atmospheric disturbances, none of which individually could have produced a storm. No single disturbance was responsible. The storm destroyed lives nonetheless.

If Glaucon's argument about one person wearing one ring was uncomfortable enough to echo across centuries of philosophy, the reality of millions of people wearing millions of rings is something closer to terrifying. The ancient philosopher asked what we become when no one is watching. The modern internet has answered his question at a scale he never dreamed was possible.

We become a swarm. We hunt in packs. We destroy with the impersonal efficiency of a natural disaster, each of us contributing our small invisible share to a force that becomes visible only in its consequences. The targets see it clearly. The rest of us look away, because the ring lets us tell ourselves that our individual contribution was too small to matter. That is the paradox at the heart of collective anonymity.

References

[1] Quinn, Zoe. *Crash Override: How Gamergate (Nearly) Destroyed My Life, and How We Can Win the Fight Against Online Hate*. PublicAffairs, 2017.

[2] Sarkeesian, Anita. Testimony before the United Nations Broadband Commission, 2015.

[3] *United States v. Tyler Barriss*, Case No. 18-CR-10048 (D. Kan. 2019).

[4] Reddit. "Reflections on the Recent Boston Crisis." Reddit blog, April 22, 2013.

[5] *R. v. Aydin Coban*, British Columbia Supreme Court, 2022.

[6] Pew Research Center. "The State of Online Harassment." January 2021.

[7] Anti-Defamation League. "Online Hate and Harassment: The American Experience." 2023.

[8] Citron, Danielle Keats. *Hate Crimes in Cyberspace*. Harvard University Press, 2014.

[9] Cyber Civil Rights Initiative. Research and statistics on non-consensual pornography.

[10] Phillips, Whitney. *This Is Why We Can't Have Nice Things: Mapping the Relationship Between Online Trolling and Mainstream Culture*. MIT Press, 2015.

Chapter Six - The Good Shepherd

Late on an October evening in 1969, a man stood alone in an office on the third floor of the RAND Corporation's Santa Monica headquarters. The building was quiet. The security guard had waved him through without question, the way he did every night, because Daniel Ellsberg was one of the most trusted analysts in American defense policy. He had the highest security clearances the government could issue. As a senior analyst, he had advised the Secretary of Defense. He had walked the battlefields of Vietnam as a civilian observer, carrying a rifle alongside Marines in the Mekong Delta. Nobody had any reason to suspect that he was about to commit what the government would call one of the most consequential acts of espionage in American history.

Ellsberg opened the combination lock on his office safe and removed a thick stack of documents. They belonged to a classified study commissioned by Secretary of Defense Robert McNamara in 1967, a forty-seven-volume analysis of United States decision-making in Vietnam from 1945 through 1967.[1] The study ran to approximately seven thousand pages. It documented, in meticulous and damning detail, that four successive presidential administrations had systematically lied to the American public about the nature, the scope, and the prospects of the war. Presidents from Truman through Johnson had known that the war was likely unwinnable. They had escalated anyway. They had concealed the escalation. At every turn, they had fabricated progress where there was none.

Ellsberg carried the documents down the hall to a Xerox machine. He began copying them one page at a time. The machine was slow. Each page took several seconds. He would return to this ritual night after night for weeks, sometimes bringing his thirteen-year-old son Robert along to help collate the pages.[1] The act was painstaking and monotonous, the kind of labor that strips away any romantic notion of what it means to defy a government. There was no dramatic confrontation, no single moment of truth. There was just a man standing beside a photocopier in an empty office, feeding page after page into the light.

He needed anonymity the way a diver needs oxygen. If his identity were discovered before the documents reached the public, the government would suppress the leak as a national security matter. The documents would vanish into classification vaults. The public would never learn what its leaders had done in its name. Ellsberg understood this calculus with perfect clarity because he had spent years inside the system that produced the lies. He had watched the machinery of deception from the inside. He knew how it responded to threats, and he knew that an identified leaker would be neutralized long before the first page reached a printing press. The ring of invisibility was not optional. It was the entire mechanism by which truth could travel from the classified world to the open one.

For nearly two years, Ellsberg tried to work through official channels. He approached members of Congress, including Senators William Fulbright and George McGovern, urging them to read the documents into the Congressional Record.[1] They declined. The political risk was too great. Ellsberg began to understand that legitimate institutions had weighed the cost of transparency and decided the public did not deserve to know. So, he made a different decision. In early 1971, he delivered copies of what would become known as the Pentagon Papers to Neil Sheehan of the New York Times.

The Times began publishing on June 13, 1971. The Nixon administration obtained a federal injunction to halt publication after three installments. It was the first time in American history that the federal government had imposed prior restraint on a major newspaper.[1] Ellsberg then provided copies to the Washington Post, and when the Post was also enjoined, he distributed copies to seventeen other newspapers in rapid succession. The Supreme Court ruled 6 to 3 in *New York Times Co. v. United States* that the government had not met the heavy burden required to justify prior restraint. The presses rolled again.

Ellsberg was identified and charged under the Espionage Act of 1917 with theft of government property and conspiracy. He faced a maximum sentence of 115 years in prison.[1] The charges were ultimately dismissed because of extraordinary government misconduct. Nixon's operatives had burglarized the office of Ellsberg's psychiatrist, searching for material to discredit him. They had wiretapped him without authorization. The judge declared a mistrial and dismissed all charges. The misconduct that saved Ellsberg was the same pattern of lawlessness that would eventually destroy Nixon's presidency.

The Pentagon Papers did not end the Vietnam War by themselves. No single document ever ends a war. But they shattered the architecture of deception that had sustained public support for the conflict. They demonstrated, with the government's own evidence, that the people had been lied to consistently and deliberately. The publication established a foundational legal precedent for press freedom under the First

Amendment. It created the framework within which future whistleblowers would operate. And it began with a man standing alone beside a photocopier in the dark, invisible long enough to get the truth out the door. The ring was not used for power, but for disclosure.

The most famous anonymous source in the history of American journalism never intended to become famous at all. For thirty-three years, the identity of the man known as Deep Throat remained one of Washington's most closely guarded secrets, a mystery that generated books, theories, accusations, and parlor games across three decades.[2] Reporters guessed. Historians speculated. Nixon's former aides pointed fingers at one another. The truth waited quietly in the mind of one man who had no intention of revealing it until he was ready.

His name was W. Mark Felt, and in 1972 he served as the Associate Director of the Federal Bureau of Investigation, the second most powerful position in the Bureau. He was fifty-eight years old, a career agent who had risen through the ranks over three decades of service. When FBI Director J. Edgar Hoover died in May of that year, Felt expected to be named his successor. The position represented the culmination of a life spent inside the institution. Nixon chose L. Patrick Gray instead, an outsider with no FBI experience whom Felt regarded as a political puppet installed to ensure White House control over the Bureau.[2] The resentment was real and it was deep. This matters because it complicates the story in a way that the heroic version prefers to ignore.

Bob Woodward, a young reporter at the Washington Post, had cultivated Felt as a source before Watergate. After five men were arrested inside Democratic National Committee headquarters at the

Watergate complex on June 17, 1972, Woodward reached out to Felt through a system of communication so elaborate it bordered on theatrical.[2] The protocols read like a spy novel. If Woodward needed to meet, he would move a flowerpot with a red flag to the balcony of his apartment. If Felt needed to initiate contact, he would circle a page number in Woodward's copy of the New York Times and draw a clock indicating the meeting time. Their meetings took place after midnight in an underground parking garage in Rosslyn, Virginia. Felt would arrive first and wait in the shadows. Woodward would take two taxis to ensure he was not followed.

The mechanics of their communication deserve attention because they illustrate something essential about the relationship between anonymity and accountability. Felt could not simply walk into the Washington Post newsroom and hand over what he knew. He was the second-ranking official of the FBI. If his identity were revealed, the investigation would shift from Nixon's crimes to Felt's violation of Bureau protocols. The information would be discredited by association with its source. Nixon's people would frame the entire Watergate story as a personal vendetta by a bitter man who had been passed over for a promotion. The anonymity was not a luxury. It was load bearing. Remove it, and the structure of accountability collapses.

Over the course of the investigation, Felt guided Woodward and Carl Bernstein toward connections that linked the Watergate burglary to the Committee to Re-Elect the President and ultimately to the White House itself.[2] He confirmed leads rather than generating them. His guidance pointed the reporters in directions they might not have pursued. He told Woodward to "follow the money," a phrase that became shorthand for investigative journalism itself, though Woodward would later note that the exact phrasing was a Hollywood invention from the 1976 film *All the President's Men*.

The outcome is well documented. The Washington Post's reporting, supported by Felt's anonymous guidance, contributed to the cascade of revelations that led to Nixon's resignation on August 9, 1974. It was the first and only resignation of an American president. Felt's identity remained concealed for more than three decades. He finally acknowledged his role in a 2005 Vanity Fair article, by which point he was ninety-one years old and suffering from dementia.[2] The revelation settled one of Washington's longest-running mysteries, but it also opened a new debate about his motivations that has never been fully resolved. The accountability mechanism that produced the most consequential journalistic investigation in American history depended entirely on one man's ability to remain invisible.

But here is the complication that the clean narrative prefers to omit. Mark Felt was not a selfless patriot acting purely in the public interest. He was angry about being passed over for the directorship. The Nixon administration's interference with the FBI's independence frustrated him deeply. He had his own institutional agenda.[2] In 1980, Felt was convicted of authorizing illegal break-ins of homes connected to the Weather Underground, a domestic terrorism group. He was pardoned by President Reagan in 1981, but the conviction stands as a reminder that the man who helped bring down a corrupt president had himself violated the rights of American citizens.

This does not diminish the importance of what Felt did. It complicates it. And the complication matters because Glaucon's thought experiment assumed a simple binary. Either the invisible man becomes a predator, or he remains a saint. Felt was neither. He was a flawed government official with mixed motives who used his invisibility to enable a genuine act of accountability. The ring does not require a pure heart to produce a just outcome. Sometimes resentment and righteousness point in the same direction, and the result is the same regardless of which one is driving. You do not have to be a saint to use the ring for good. You just have to aim it at something that deserves the light.

On May 20, 2013, Edward Snowden boarded a flight from Honolulu to Hong Kong carrying four laptop computers loaded with thousands of classified documents from the National Security Agency.[3] He was twenty-nine years old. He had worked as a systems administrator for the NSA through contracting firms Booz Allen Hamilton and Dell. His security clearance gave him access to some of the most sensitive surveillance programs in the United States intelligence apparatus. He had decided, over a period of months, that the American public had a right to know what those programs were doing.

The operational security he maintained during the disclosure process was extraordinary. He had established encrypted communications with filmmaker Laura Poitras and journalist Glenn Greenwald months before leaving the country. His primary tool was the Tails operating system, which routes all internet traffic through the Tor anonymity network, leaving no trace on the machine being used.[4] He communicated through encrypted email and required Greenwald to set up PGP encryption before he would share any details about the material he possessed. When Greenwald was slow to set up the encryption, Snowden grew frustrated but did not waver. He understood that the security infrastructure was not paranoia. It was the architecture that would determine whether the documents reached the public or disappeared into a prosecution.[3]

Snowden chose Hong Kong for the initial disclosure because it sat in a legal gray zone, outside the immediate reach of American law enforcement but connected enough to the global media infrastructure that journalists could reach him quickly. He met Greenwald, Poitras, and Guardian reporter Ewen MacAskill in a hotel room at the Mira Hotel in Kowloon. Before saying a word, he asked them to place their cell phones in the hotel refrigerator to block any potential signals. Then he began explaining what he had brought.[3]

The revelations came in waves. The first stories, published in the Guardian and the Washington Post in June 2013, revealed the existence of PRISM, a program through which the NSA collected data directly from the servers of nine major technology companies including Google, Facebook, Apple, and Microsoft.[3] Subsequent disclosures revealed XKeyscore, a system that allowed analysts to search through vast databases of emails, browsing histories, and chat logs with virtually no oversight. They revealed bulk collection of telephone metadata from millions of American citizens under a secret interpretation of Section 215 of the Patriot Act. Other documents showed that the NSA had tapped the personal cell phone of German Chancellor Angela Merkel.[3] Still others revealed that the agency had worked to weaken encryption standards, the very tools that ordinary people relied on to protect their private communications.

The public learned, in document after document, that its government had built the most comprehensive surveillance apparatus in human history. It had done so in secret, under the authority of a court whose proceedings were classified, with the cooperation of technology companies that were legally prohibited from acknowledging their participation.

Snowden was charged under the Espionage Act of 1917, the same statute used against Daniel Ellsberg four decades earlier.[4] He left Hong Kong for Moscow, where he was granted asylum by the Russian government. For years he lived in Russia, eventually receiving Russian citizenship. The circumstances of his exile provided ammunition to critics who accused him of being a Russian asset or a traitor, though no evidence has been produced to support either claim.[4]

The debate over Snowden's actions has never been resolved, and perhaps it cannot be. The intelligence community argues that his disclosures damaged national security and endangered lives. Former NSA Director Keith Alexander claimed the leaks caused "significant and irreversible damage" to intelligence capabilities.[3] Civil liberties organizations argue that the programs he revealed were unconstitutional and that the public had a right to know about them. Federal courts have since ruled that the bulk metadata collection program was illegal. Congress passed the USA Freedom Act in 2015 to curtail some of the surveillance practices Snowden exposed. Technology companies redesigned their encryption protocols in response to the revelations, making end-to-end encryption a standard feature rather than a specialist tool.[4] The debate over whether he is a hero or a traitor is an ongoing topic.

The story of Reality Winner illustrates what happens when the ring fails. In May 2017, Winner, a twenty-five-year-old NSA contractor and Air Force veteran, printed a single classified document and mailed it to the online news outlet The Intercept.[5] The document detailed Russian intelligence efforts to hack American election infrastructure in the lead-up to the 2016 presidential election. Winner believed the public needed to see evidence that a foreign government had attacked American democracy.

Her anonymity lasted five days. When The Intercept contacted the NSA for comment before publication, the agency examined the document and identified microscopic tracking dots embedded by the printer that linked the printout to a specific machine.[5] Investigators determined that only six people had printed the document. Of those six, only Winner had been in email contact with the news outlet. She was arrested at her home in Augusta, Georgia, on June 3, 2017.

Winner was charged under the Espionage Act and pleaded guilty to one count of unauthorized transmission of national defense information. She was sentenced to five years and three months in federal prison, the longest sentence ever imposed for an unauthorized disclosure to the media at that time.[5] She served more than four years before being released to a halfway house in 2021.

The contrast between Snowden and Winner is instructive and deeply unsettling. Snowden's elaborate operational security allowed him to reach journalists, transfer the documents, and leave the country before the government could respond. He had spent months studying the architecture of surveillance and building countermeasures against it. Winner's disclosure involved a single printed page, no encryption, no anonymizing tools, and a news outlet that inadvertently helped identify her by sharing the original document with the very agency she had leaked it from. The ring of invisibility is not a metaphor for these people. It is an operational reality. When it holds, truth reaches the public. When it breaks, the person holding the truth goes to prison. The stakes are not abstract.

The tools that Snowden used to protect himself did not materialize from the private sector. They were born inside the United States military. In the mid-1990s, mathematicians at the U.S. Naval Research Laboratory began developing a system called onion routing, a technique for anonymizing internet communications by encrypting data in multiple layers and routing it through a series of volunteer-operated servers around the world.[6] Each server peels away one layer of encryption, learning only the identity of the previous server and the next one, never the origin or the destination. The system became the Tor Project, released as open-source software in 2003 and funded in its early years by the Defense Advanced Research Projects Agency and the State Department.

The irony is sharp enough to cut. The United States government funded the creation of the most powerful anonymity tool on the internet. It did so because American intelligence agencies needed a way to communicate covertly online, and a tool that only intelligence agents use is not anonymous at all. If every Tor user is a spy, then using Tor identifies you as a spy. For the system to work, it needed a large civilian user base to provide cover traffic. The government needed ordinary people to use the system so that intelligence traffic could hide in the crowd.[6] The ring of invisibility was distributed to the public not as a gift but as camouflage for the powerful. The public was meant to serve as noise in which the signal could hide. What the public did with it exceeded any scenario the Naval Research Laboratory had imagined. The ring they built for spies became, in the hands of ordinary citizens, something far more powerful than a tool for covert operations. It became the infrastructure of dissent.

In December 2010, a twenty-six-year-old Tunisian street vendor named Mohamed Bouazizi set himself on fire outside a government building in the city of Sidi Bouzid. His self-immolation, a protest against police corruption and economic desperation, was recorded on a cell phone and uploaded to Facebook. Within days, protests spread across Tunisia. Within weeks, they reached Egypt, Libya, Bahrain, Syria, and Yemen. The movement that came to be known as the Arab Spring had begun.[8]

The protesters faced governments that controlled the telecommunications infrastructure, monitored internet traffic, and had demonstrated a willingness to imprison or kill dissidents. Tunisian activists used Tor to access blocked websites and to communicate with foreign journalists without revealing their identities to the Ben Ali regime's security services. Egyptian activists used Virtual Private Networks and encrypted messaging tools to coordinate protests while circumventing government censorship.[8] When the Mubarak regime shut down Egypt's internet entirely on January 28, 2011, activists found workarounds through dial-up connections, satellite links, and ham radio operators who relayed messages to the outside world. The anonymity tools did not cause the revolution. Decades of repression caused the revolution. But the tools allowed organizers to function in an environment designed to make their work fatal.

The pattern repeats across authoritarian states. In China, citizens use Tor and VPN services to circumvent the Great Firewall, the government's comprehensive system of internet censorship that blocks access to foreign news outlets, social media platforms, and any content the Communist Party considers threatening.[6] The scale of what is hidden from Chinese citizens is staggering. Information about the Tiananmen Square massacre, the detention of Uyghur Muslims in Xinjiang, the activities of exiled dissidents, and the internal politics of the Party leadership exists behind a wall of censorship that Tor allows some citizens to breach. For those users, the ring of invisibility is not a tool for mischief. It is the only mechanism through which they can access information that their government has decided they should not see.

In countries where homosexuality is criminalized, anonymous platforms provide connections that could not safely exist in the open. LGBTQ+ individuals in Uganda, Iran, Saudi Arabia, and dozens of other nations use encrypted messaging and anonymous browsing to access community resources, health information, and human contact.[6] In some of these countries, the penalty for being identified is imprisonment. In others, it is death. The ring of invisibility is not a philosophical abstraction for a gay teenager in Tehran. It is the difference between connection and absolute isolation, between safety and a prison cell.

The situation for journalists in Mexico reveals a particularly grim dimension of this calculus. Mexico has been one of the deadliest countries in the world for reporters for more than a decade. According to the Committee to Protect Journalists, more than 150 journalists and media workers have been killed in Mexico since 1992, with the vast majority of cases linked to reporting on organized crime and government corruption.[7] Reporters Without Borders consistently ranks Mexico among the most dangerous countries for press freedom.[9] In states like Tamaulipas, Guerrero, and Sinaloa, cartel violence has created zones where no journalist can safely report under their own name.

The most important investigative reporting on cartel activity, government collusion, and mass disappearances often comes from anonymous sources, anonymous blogs, and anonymous social media accounts operated by citizens who understand that attaching their names to the truth would result in their deaths.[7] The anonymous Twitter accounts and Facebook pages that document cartel violence in Mexican border cities exist because their operators have watched what happens to people who do the same work publicly. They have attended the funerals. They have seen the bodies. The ring is not a convenience for them. It is survival equipment.

The infrastructure of safe anonymity extends beyond any single tool. Signal, the encrypted messaging application, provides end-to-end encryption that prevents anyone except the sender and recipient from reading the contents of a message. ProtonMail offers encrypted email services based in Switzerland, beyond the reach of many national surveillance systems. SecureDrop, developed by Aaron Swartz and later maintained by the Freedom of the Press Foundation, allows whistleblowers to submit documents to news organizations through an anonymized channel.[6] Each of these tools represents a deliberate engineering decision to build systems in which the ring of invisibility is a feature rather than a flaw.

The argument is not complicated. For people living under authoritarian regimes, in conflict zones, in countries where speaking the truth carries a death sentence, the ring of invisibility is not a moral luxury. It is a precondition for the exercise of basic human rights. Freedom of speech means nothing if speaking can be traced to the speaker. The right of assembly means nothing if assembling can be monitored by the state. Freedom of the press means nothing if sources can be identified and punished. These are not hypothetical concerns. They describe the daily reality for billions of people living under governments that treat information as a threat. The ring makes those freedoms operational in environments designed to suppress them.

Now we arrive at the problem that has no solution. It is the problem that every legislator, every platform designer, every civil liberties lawyer, and every intelligence official confronts when they try to make policy about anonymity in the digital age. It is a problem that Plato never imagined because he never conceived of a world in which the ring would be mass-produced.

The previous chapter documented what the ring of invisibility produces when it is worn by people who want to destroy. Coordinated harassment campaigns, the targeted annihilation of individual lives, mob justice visited upon the innocent, the slow grinding erosion of public discourse under the weight of consequence-free cruelty. That chapter told a true story. Every case was real. The harm was measurable and documented.

This chapter has told a different true story. Whistleblowers who exposed government deception. Anonymous sources who enabled the only presidential resignation in American history. A contractor who revealed that his government was surveilling its own citizens on a scale previously unimaginable. Dissidents who organized revolutions against dictatorships. Journalists who reported on cartels from behind the shield of anonymity because the alternative was death. Citizens who accessed forbidden information through tools designed to make them invisible. Every case in this chapter is also real. The courage is also measurable, even if it is harder to quantify than the harm.

The paradox sits at the center of every policy debate about anonymity, encryption, and digital privacy, and it is this... the same technology that enabled Gamergate also enabled the Arab Spring. The same Tor network that allows a harasser to threaten a woman with impunity also allows a dissident in Damascus to communicate with a journalist in London without being arrested and tortured. The same encryption that shields a child predator from law enforcement also shields a whistleblower from retaliation. The infrastructure does not distinguish between its users. It cannot examine intentions. It provides the ring to everyone who asks for it and makes no judgment about what they will do once they become invisible.

Every proposal to weaken anonymity in order to prevent abuse simultaneously weakens the protection that vulnerable people depend on. When governments demand backdoors in encryption, they are asking for a mechanism that can be used by intelligence agencies in democratic countries but also by security services in authoritarian ones. Platforms that implement mandatory real-name policies reduce harassment in some contexts but make participation dangerous for dissidents, activists, and marginalized people in others. When legislators propose that anonymous speech should carry legal liability, they create a framework that applies equally to the troll and the whistleblower.

The Australian government learned this in 2021 when it passed legislation requiring technology companies to provide law enforcement access to encrypted communications. Civil liberties organizations immediately noted that the same capabilities would be available to any government that obtained the cooperation of those companies, including governments with records of persecuting journalists and political opponents.[6] The tool built to catch criminals becomes, in different hands, a tool to catch dissidents. There is no technical mechanism that limits a backdoor to virtuous users.

Glaucon never considered this possibility. In Book II of the Republic, he presented the ring as a test of individual character. A man becomes invisible. Does he remain just? The entire framing assumes that invisibility serves only the wearer. It assumes that the only question worth asking is whether the invisible person will abuse the power.

But the evidence of the past fifty years suggests a more complicated reality. Some people, given invisibility, become predators. They harass, they threaten, they destroy. The previous chapters documented this with precision. Other people, given the same invisibility, become protectors. They expose corruption, enable accountability and provide a lifeline to the persecuted. This chapter has documented that with equal precision.

The uncomfortable truth is that you cannot design a system that grants the ring only to those who will use it well. Every attempt to sort the righteous from the wicked requires a sorting mechanism, and every sorting mechanism requires surveillance. The act of determining who deserves anonymity destroys the anonymity in the process of evaluating it. You cannot ask someone to prove they are worthy of invisibility without first making them visible.

This is not a failure of engineering. It is a structural feature of the problem. The ring cannot be programmed to recognize the hand that wears it. It amplifies whatever the wearer brings to it. Some people use the ring to take. Others use it to give. The majority, as we will see in later chapters, do neither. They simply carry on with their lives, neither saints nor predators, their moral character unaltered by the presence or absence of an audience.

The question, then, has never really been about the ring. Plato was closer to the truth than he may have realized. The ring is a tool, as morally inert as a hammer or a printing press or a network router. The question is about the hand that wears it. It is about what we build into people before they put it on. It is about whether a society can produce citizens who will choose justice even when injustice is free, who will protect the vulnerable even when cruelty is costless, who will tell the truth even when lies are safe.

References

[1] Ellsberg, Daniel. *Secrets: A Memoir of Vietnam and the Pentagon Papers*. Viking, 2002.

[2] Woodward, Bob. *The Secret Man: The Story of Watergate's Deep Throat*. Simon & Schuster, 2005.

[3] Greenwald, Glenn. *No Place to Hide: Edward Snowden, the NSA, and the U.S. Surveillance State*. Metropolitan Books, 2014.

[4] Harding, Luke. *The Snowden Files: The Inside Story of the World's Most Wanted Man*. Vintage Books, 2014.

[5] *United States v. Reality Leigh Winner*, Case No. 1:17-CR-034 (S.D. Ga. 2017).

[6] Dingledine, Roger, Nick Mathewson, and Paul Syverson. "Tor: The Second-Generation Onion Router." *USENIX Security Symposium*, 2004.

[7] Committee to Protect Journalists. "Killed in Mexico." CPJ data, updated annually.

[8] Howard, Philip N., and Muzammil M. Hussain. *Democracy's Fourth Wave? Digital Media and the Arab Spring*. Oxford University Press, 2013.

[9] Reporters Without Borders. *World Press Freedom Index*. Published annually.

Chapter Seven - The Invention of Invisible Money

The river Pactolus still flows through what was once the kingdom of Lydia. It descends from the slopes of Mount Tmolus, cuts through the hills of western Turkey, and winds past the ruins of Sardis toward the broad coastal plain of the Aegean. Tourists who visit the site today find a modest stream running through an archaeological park where the remains of a Roman gymnasium sit beside the foundations of a Lydian marketplace. The water runs clear now. Twenty-seven centuries ago it carried something far more valuable than fish or irrigation. The Pactolus carried gold.

The ancient Greeks believed King Midas had washed his golden touch away in this river, and that the flecks of precious metal in its silt were remnants of that curse lifted. The Lydians had a more practical explanation. The river deposited a natural alloy of gold and silver called electrum along its banks, and the people of Sardis learned to collect it. They learned to smelt it. They learned to stamp it into uniform lumps bearing the seal of their king. Those lumps became the first coins in recorded history.[1]

The dating remains a subject of scholarly debate. Most numismatists place the first Lydian coins somewhere between 630 and 600 BCE, during a period when Lydia was ascending toward regional dominance under the Mermnad dynasty.[2] The coins were small and bean-shaped, stamped on one side with the image of a roaring lion. They weighed a consistent amount. They bore an official mark guaranteeing their value. These two features transformed them from mere metal into something unprecedented. They were standardized, portable units of wealth that could move from hand to hand without any record of who had held them before.

Before coins existed, trade required presence. A farmer who wanted cloth had to find a weaver who wanted grain, and the two of them had to stand face to face and negotiate. The transaction demanded mutual knowledge. Each party assessed the other's reliability, weighed the quality of the goods, and calculated the fairness of the exchange in real time. Barter transactions were personal by nature. Both parties knew each other. Both parties witnessed the exchange. Neighbors witnessed it too. The community itself served as a ledger, because reputation was the only currency that truly mattered. A man known for cheating would find his trading partners disappearing. A man known for fair dealing would attract them. Commerce was a social act embedded in relationships, and those relationships enforced a rough honesty that no written law could have replicated.

Coins changed everything. A merchant in Sardis could accept a stamped piece of electrum from a stranger, knowing that the metal held its value regardless of the stranger's name. The coin did not care who carried it. The coin did not remember where it had been. A thief could spend a stolen coin as easily as an honest laborer could spend an earned one, and the marketplace would treat both transactions identically. The abstraction was revolutionary. For the first time in human history, value could travel anonymously. Wealth could accumulate in silence. Power could be acquired without the community's knowledge or consent.

The kingdom that produced this invention was the same kingdom that produced the story of the shepherd who found a ring of invisibility and used it to seize power. Gyges of Lydia, whether historical king or philosophical thought experiment, belonged to the same culture that gave humanity its first technology for making wealth invisible.[1] This was not a coincidence. It was a pattern that would repeat itself across time. Anonymity and financial power have been linked from the very beginning, because money is itself a ring of invisibility. It separates action from consequence. It separates identity from transaction. It allows the holder to move through the world without leaving a trace.

The centuries after Lydia produced a long chain of innovations, and each one made financial invisibility more sophisticated. The progression was neither accidental nor conspiratorial. It followed a simple logic. Wherever people accumulated wealth, they sought ways to protect it from scrutiny. Wherever governments imposed taxes or restrictions, financial engineers invented new instruments to evade them. The tools changed with the times, but the underlying impulse remained constant. People with money wanted to move it without being watched.

Bearer bonds were among the earliest modern instruments of anonymous wealth. The concept was elegantly simple. A bearer bond was a physical certificate that entitled whoever held it to receive interest payments and the return of principal at maturity. The bondholder's name appeared nowhere on the document. Possession was ownership. If a man lost a bearer bond on the street and another man picked it up, the second man became the rightful owner. Bearer bonds turned wealth into something as anonymous as cash but far more portable in large amounts. A single piece of paper could represent millions of dollars, and it could change hands with no record of the transfer.[3]

Governments issued bearer bonds for decades. Corporations issued them freely. They circulated through the financial system like ghosts, carrying value without identity. By the mid-twentieth century, the potential for abuse had become obvious. Tax evaders used bearer bonds to hide income. Money launderers used them to clean dirty cash. Criminal organizations used them to move funds across borders without detection. The United States effectively ended domestic issuance of bearer bonds through the Tax Equity and Fiscal Responsibility Act of 1982, but by then the concept had already migrated into more sophisticated forms.[4]

Switzerland formalized the next great leap in financial secrecy with the Swiss Banking Act of 1934. The law made it a criminal offense for a Swiss banker to reveal the identity of a client to any third party, including foreign governments. The numbered account became the symbol of this system. A depositor could open an account identified only by a number, with the connection between the number and the depositor's true identity known to just a handful of senior bankers.[5]

The origins of this law carried a particular moral weight. In the early 1930s, Nazi agents had been crossing into Switzerland to identify German Jews who held Swiss bank accounts. The Nazis wanted to seize those assets, and several cases of exposure had already led to confiscations and worse. The Swiss Banking Act was designed in part to protect vulnerable depositors from a predatory government. It was a shield built for the persecuted.[5]

Within a generation, the shield had become a sword. Dictators, arms dealers, drug traffickers, and corrupt politicians from every continent discovered that Swiss banking secrecy offered them the same protections it had offered Jewish refugees. Ferdinand Marcos of the Philippines stashed billions in Swiss accounts. Mobutu Sese Seko of Zaire did the same. The Duvalier family of Haiti followed suit. Swiss banks did not ask where the money came from. Swiss law made it a crime for them to tell anyone it was there. The tool that had been created to protect the innocent became the preferred instrument of the guilty.[5]

Shell companies represented the next evolution. A shell company is a legal entity that exists on paper but conducts no actual business operations. It can own assets, hold bank accounts, and enter into contracts. What it cannot easily reveal is who controls it. The concept dates to the nineteenth century, when entrepreneurs discovered they could create limited liability companies to hold property and investments while keeping their own names out of public records.[6] By the twentieth century, shell companies had become the fundamental building blocks of financial invisibility.

The mechanics were straightforward. A lawyer in Panama or the British Virgin Islands could incorporate a company in a matter of hours, listing nominee directors whose names appeared on the paperwork in place of the true owner. That company could then open bank accounts, purchase real estate, and conduct financial transactions under its corporate name. If anyone investigated, they would find only the shell. The beneficial owner remained invisible behind layers of corporate structure.

The sophistication of these arrangements grew over time. A single wealthy individual might establish a holding company in the British Virgin Islands that owned a subsidiary in Panama, which in turn held accounts in Switzerland and Luxembourg. The layers served no business purpose.

They existed solely to create distance between the owner and the assets, to make the chain of ownership so long and so tangled that no investigator could follow it to its end. Lawyers who specialized in these structures called the practice "layering," and they charged handsomely for it.[6]

Offshore jurisdictions built entire economies around this service. The British Virgin Islands, with a population smaller than many American suburbs, became home to more than 400,000 registered companies.[7] The Cayman Islands, a British overseas territory in the Caribbean, accumulated financial assets worth more than sixty times its annual GDP. Panama, Luxembourg, Jersey, Guernsey, the Isle of Man, Liechtenstein... the list of jurisdictions competing to offer the deepest secrecy grew longer with each passing decade. These were not rogue states operating outside the international system. Many of them were territories of major Western democracies, operating with the tacit approval of London and Washington.[7]

The arrival of cryptocurrency in 2008 represented the latest chapter in this long history. The timing was significant. In September 2008, the global financial system nearly collapsed. Lehman Brothers filed for bankruptcy. Governments spent trillions of taxpayer dollars bailing out banks whose reckless behavior had caused the crisis. Trust in financial institutions fell to historic lows. On October 31 of that year, just weeks after the worst of the panic, an individual or group operating under the pseudonym Satoshi Nakamoto published a nine-page whitepaper titled "Bitcoin: A Peer-to-Peer Electronic Cash System."[8] The paper proposed a digital currency that would operate without banks, without governments, and without any central authority. Transactions would be verified by a decentralized network of computers and recorded on a public ledger called a blockchain. The vision was explicitly anti-institutional. Bitcoin was born from a crisis of trust, and it promised to replace trust with mathematics.

The identity of Satoshi Nakamoto has never been confirmed. The creator of the world's most famous financial technology chose to remain anonymous. The irony was almost too precise to be accidental. The inventor of a system designed to facilitate pseudonymous transactions operated under a pseudonym. The ring of invisibility had found its most eloquent advocate in a person who did not exist.[8]

Bitcoin itself turned out to be pseudonymous rather than truly anonymous. Every transaction was recorded on the public blockchain, and sophisticated analysts could sometimes trace the flow of funds back to identifiable individuals. Law enforcement agencies developed tools to follow the digital trail. But the cryptocurrency ecosystem quickly produced alternatives designed to close even these gaps. Monero, launched in 2014, used ring signatures and stealth addresses to make transactions untraceable. Zcash offered a protocol called zk-SNARKs that allowed transactions to be verified without revealing the sender, the receiver, or the amount. Each new privacy coin promised what its predecessors had failed to deliver... complete financial invisibility in the digital age.

The pattern held across every innovation. Bearer bonds were created to facilitate legitimate capital markets. Swiss bank accounts were created to protect the persecuted. Shell companies were created to manage legitimate business complexity. Cryptocurrency was created to decentralize finance and reduce dependence on institutions that had failed during the 2008 financial crisis. Each tool served a genuine purpose. Each tool was then exploited by those who sought to act without being seen.

On a Sunday evening in the spring of 2015, a reporter at the Suddeutsche Zeitung in Munich received an encrypted message from an unknown source. The message arrived through a secure channel and contained no identifying information about the sender. The source identified themselves only as "John Doe" and asked a single question. The source wanted to know if the newspaper was interested in receiving data.[9]

The reporter was Bastian Obermayer, a veteran investigative journalist who specialized in financial crime. He had received tips from anonymous sources before, and he understood immediately that such an offer could mean anything from a minor corporate scandal to an elaborate hoax designed to waste his time or feed him disinformation. He responded carefully, establishing ground rules for communication. He asked what kind of data. The answer came back in fragments over the following days, each message encrypted and each one revealing a little more about the scope of what the source possessed.

The source claimed to possess internal documents from a law firm in Panama called Mossack Fonseca. The source claimed the documents revealed a global network of hidden wealth involving politicians, criminals, and billionaires on every continent. The source claimed the archive was enormous.[9]

Obermayer and his colleague Frederik Obermaier began reviewing the first batches of material. What they found exceeded anything they had imagined. Mossack Fonseca was one of the world's largest providers of offshore corporate services. For decades, the firm had helped clients create shell companies in jurisdictions around the world. The leaked documents contained internal emails, financial records, corporate registrations, and passport scans spanning forty years of operations. The total volume was staggering. It amounted to 2.6 terabytes of data, encompassing 11.5 million individual documents.[10] It was the largest leak in journalistic history at that time.

The Suddeutsche Zeitung recognized that no single newsroom could process this volume of material. The reporters brought the data to the International Consortium of Investigative Journalists in Washington, DC, an organization built specifically for cross-border investigative collaborations. The ICIJ coordinated a partnership that would eventually involve nearly four hundred journalists from 107 media organizations across 80 countries.[11] The scale of the collaboration was itself unprecedented. Reporters from Le Monde in Paris, The Guardian in London, the Indian Express in New Delhi, and dozens of other outlets shared data and findings on a secure platform accessible only to vetted participants. For over a year, these journalists worked in secret, analyzing the documents, cross-referencing names against public records, and building stories that connected the offshore entities to real people with real power. They communicated through encrypted channels and operated under strict security protocols. The secrecy was essential, because the people named in the documents included heads of state, intelligence chiefs, and billionaires with the resources to suppress stories before publication.

On April 3, 2016, the ICIJ published the Panama Papers.[11] The revelations were explosive. The documents exposed 214,000 offshore entities linked to people in more than 200 countries and territories. The names included 140 politicians and public officials from 50 countries.[11] The scale was almost incomprehensible. A single law firm in a single Central American country had served as the invisible hand behind hundreds of thousands of financial transactions designed to hide wealth from public view.

The stories that emerged read like a catalog of the world's most powerful people caught operating in the dark. The details were specific, the numbers were precise, and the names belonged to people who had built their public identities on images of trustworthiness and service. In Iceland, Prime Minister Sigmundur Gunnlaugsson and his wife owned a company called Wintris, registered in the British Virgin Islands through Mossack Fonseca. The company held millions of dollars in bonds from Iceland's failed banks. Gunnlaugsson had participated in political negotiations over those same bank failures without disclosing his personal financial interest. When the Panama Papers revealed the connection, thousands of Icelanders gathered outside the parliament building in Reykjavik. Gunnlaugsson resigned within days.[11]

The documents traced approximately $2 billion through a network of shell companies and bank accounts connected to associates of Russian President Vladimir Putin. The central figure in this network was Sergei Roldugin, a classical cellist who happened to be Putin's close childhood friend. Roldugin's official income could not begin to explain the vast sums flowing through companies linked to his name. The shell companies had moved money through banks in Switzerland, Cyprus, and other offshore centers, generating profits and distributing funds in patterns that investigators found consistent with a system designed to hold wealth on behalf of someone else.[9] Putin denied any wrongdoing. Roldugin told journalists that the money had been used to purchase rare musical instruments.

The breadth of the revelations defied any single narrative. The documents implicated associates of the King of Saudi Arabia, the President of Ukraine, and the President of Argentina. They exposed the Mack family in Argentina, wealthy business figures in China, prominent names in India and Pakistan, and football stars whose agents had structured their image-rights payments through offshore entities. The Panama Papers did not reveal a single conspiracy. They revealed a system, an entire architecture of financial invisibility that operated in plain sight for anyone wealthy enough to access it.

The aftermath was uneven. Some countries launched investigations. Some politicians resigned or faced prosecution. Iceland's prime minister lost his office. Pakistan's Prime Minister Nawaz Sharif was eventually disqualified from holding office by the country's Supreme Court, partly on the basis of undisclosed assets revealed in the Papers.[11] Mossack Fonseca itself shut down in 2018, though its founders faced legal proceedings.

But many of the people named in the documents faced no consequences at all. The fundamental problem was one of legality. In many jurisdictions, the offshore structures revealed by the Panama Papers were technically legal. Owning a shell company was not a crime. Holding money overseas was not a crime. The line between aggressive tax avoidance and criminal tax evasion was blurry enough for expensive lawyers to defend, and the people implicated in the Papers could afford the most expensive lawyers in the world. Prosecutors in countries with underfunded justice systems found themselves outmatched by legal teams whose billable hours exceeded the annual budgets of the agencies pursuing them. The architecture of invisibility had been built within the law, and the law proved inadequate to dismantle it.

John Doe, the anonymous source who had started it all, eventually published a manifesto explaining the leak. The document described income inequality as "one of the defining issues of our time" and called for the end of the offshore system.[9] The source never revealed their identity. The person who had made the data visible chose to remain invisible themselves, a final irony in a story already saturated with them.

Five years after the Panama Papers, the architecture proved itself to be not merely resilient but expanding. On October 3, 2021, the ICIJ published the Pandora Papers.[12] The new investigation drew on 11.9 million documents from 14 different offshore service providers. The scope was even broader than the Panama Papers. The findings were, in many ways, worse, because they demonstrated that the first exposure had changed almost nothing.

King Abdullah II of Jordan had used shell companies to accumulate more than $100 million worth of luxury real estate across the United States and the United Kingdom.[12] The purchases included oceanfront properties in Malibu, a Georgetown mansion in Washington DC, and apartments in central London. All of these assets were held through a web of offshore entities registered in the British Virgin Islands and other jurisdictions, with each layer of corporate ownership designed to conceal the king's connection to the properties. Jordan's royal court issued a statement saying the properties were being held for security and privacy reasons, and that no public funds had been used. The explanation satisfied few of the king's subjects in a country where unemployment hovered near 25 percent and public services were chronically underfunded.

Czech Prime Minister Andrej Babis had purchased a $22 million chateau in the south of France through a chain of shell companies routed through the British Virgin Islands, Monaco, and Washington DC.[12] The purchase occurred in 2009, years before Babis entered politics. But the structure that concealed his ownership remained active even as he campaigned for office on an explicit anti-corruption platform. He promised Czech voters that he would fight the culture of hidden money and insider dealing. The Pandora Papers revealed that he had been a participant in exactly that culture.

The documents named hundreds of other politicians from countries around the world, including leaders who had specifically promised reform in the wake of the Panama Papers. The President of Kenya, the Prime Minister of Pakistan, the President of Ecuador, the King of Morocco, and dozens of other heads of state or senior officials from nations on every continent appeared in the files.[12] Some of them had come to power on platforms explicitly promising to combat corruption. Some had signed international agreements committing their governments to financial transparency. The same dynamic repeated without variation. Wealth flowed outward through shell companies, settled in offshore jurisdictions, and returned to the beneficial owners through mechanisms designed to prevent public scrutiny. The public promises meant nothing when the private channels remained open.

The Pandora Papers also revealed something the Panama Papers had only hinted at. The United States itself had become a major haven for hidden wealth. The state of South Dakota had attracted tens of billions of dollars into trust companies that offered levels of secrecy rivaling anything in the Caribbean. Delaware had long been the preferred jurisdiction for corporate formation in America, with more than a million business entities registered in a state with fewer than a million residents.[13] The architects of financial invisibility did not need to travel to exotic islands. They could find everything they needed in the American heartland.

The lesson of the Pandora Papers was stark. Shell companies functioned as the corporate ring of Gyges. They allowed individuals to act in the financial world without their names attached. They allowed the powerful to accumulate wealth, purchase property, and move money across borders while remaining invisible to the public, to regulators, and to the tax authorities of their own countries. The first great exposure had provoked outrage, headlines, and a handful of resignations. The system itself had barely flinched.

Daphne Caruana Galizia understood the architecture of invisible money better than almost any journalist alive. She had spent years mapping its contours in Malta, a small island nation in the Mediterranean that had positioned itself as a bridge between European respectability and offshore secrecy. Malta was barely visible on most maps of Europe. Its total land area was smaller than most American cities. Its population numbered fewer than half a million people. But its financial services sector punched far above its weight, attracting investment and corporate registrations from across the globe. The Maltese government had deliberately cultivated this industry, offering favorable tax treatment and a passport-by-investment program that granted European Union citizenship to wealthy foreigners in exchange for substantial financial commitments. Caruana Galizia saw the corruption that festered beneath this arrangement. She wrote about it relentlessly.[14]

Her blog, Running Commentary, became the most widely read publication in Malta. She published multiple times a day, exposing the connections between Maltese politicians, business leaders, and the offshore structures they used to conceal their dealings. Her reporting was fearless in its specificity. She named names. She published documents. She traced the flow of money through shell companies registered in Panama, the British Virgin Islands, and other offshore jurisdictions.[14]

When the Panama Papers broke in 2016, Caruana Galizia was among the Maltese journalists who received access to the documents through the ICIJ collaboration. What she found confirmed her darkest suspicions. The documents revealed that two senior members of Malta's government, the energy minister Konrad Mizzi and the prime minister's chief of staff Keith Schembri, had set up shell companies in Panama through Mossack Fonseca.[14] The companies were established shortly after both men took office, and they appeared to be connected to a Dubai-based company called 17 Black. Caruana Galizia reported that 17 Black was set to receive payments from a consortium involved in Malta's energy sector.

The revelations should have brought down the government. In most functioning democracies, the exposure of senior officials hiding offshore companies would have triggered resignations and criminal investigations.

In Malta, the reaction was different. Prime Minister Joseph Muscat refused to dismiss the implicated officials. The police declined to pursue investigations. The political establishment closed ranks.[14]

Caruana Galizia faced an escalating campaign of intimidation. She received death threats by phone, by email, and through intermediaries. Her family's front door was set on fire. Her dogs were poisoned. Her son Matthew found the family's mailbox destroyed. She was subjected to more than forty libel suits filed by powerful individuals seeking to silence her through legal and financial pressure, a strategy known as SLAPP litigation that was designed to drain her resources rather than to win in court.[15] She documented each threat on her blog with the precision of a forensic accountant tallying entries in a ledger. She continued publishing.

On the afternoon of October 16, 2017, Daphne Caruana Galizia left her home in Bidnija, a village in the north of Malta. She got into her car, a white Peugeot 108, and drove toward the main road. Approximately 280 meters from her home, a bomb detonated beneath the vehicle.[15] The blast was powerful enough to scatter wreckage across a wide area. Caruana Galizia was killed instantly. She was fifty-three years old.

The murder sent shockwaves through Malta and across the international journalism community. Within hours, crowds began gathering in Valletta, Malta's capital. Thousands marched through the ancient streets carrying photographs of Caruana Galizia and signs demanding justice. Vigils sprang up at a makeshift memorial outside the Palace of Justice, and these vigils continued for months, becoming a permanent expression of public grief and anger. The European Parliament passed a resolution calling for an independent investigation.

Reporters Without Borders and the Committee to Protect Journalists condemned the assassination. Press freedom organizations around the world pointed to the failures of the Maltese state to protect a journalist who had been living under constant and credible threat for years.[15]

The investigation that followed was long, contested, and deeply revealing of the forces that had aligned against Caruana Galizia during her life. For months, progress was agonizingly slow. Critics accused the Maltese police of dragging their feet, of protecting the powerful interests that the journalist had spent her career exposing. International pressure eventually forced movement. Three men were arrested and charged with carrying out the bombing. Vince Muscat, Alfred Degiorgio, and George Degiorgio were accused of planting the explosive device that had been attached to the underside of her car. Muscat later pleaded guilty and cooperated with investigators, providing crucial testimony about who had commissioned the killing.[15]

In November 2019, Maltese police arrested Yorgen Fenech, one of Malta's wealthiest businessmen, as he attempted to leave the country on his yacht. Fenech was charged with complicity in the murder. He was identified as the owner of 17 Black, the Dubai company that Caruana Galizia had linked to the government officials.[15] The connection was devastating in its clarity. The journalist had been investigating the flow of money from Fenech's business interests through offshore structures to politicians in the Maltese government. She had been killed for making invisible money visible.

A public inquiry commissioned by the Maltese government concluded in 2021 that the state bore responsibility for creating an atmosphere of impunity that led to Caruana Galizia's assassination.[15] The inquiry found that the government had failed to recognize the real and immediate risk to her life. It found that the culture of corruption she had exposed was enabled by the state itself. The report was damning in its assessment that Malta's political leadership had prioritized the protection of powerful interests over the protection of a journalist whose work threatened to expose them.

Prime Minister Joseph Muscat had resigned in December 2019 amid the fallout from the investigation. Konrad Mizzi resigned from parliament. Keith Schembri was arrested on corruption charges. The machinery of accountability ground forward with excruciating slowness, producing consequences that arrived years after the crime.

Daphne Caruana Galizia's sons carried on her legacy with a determination that honored her example. Matthew Caruana Galizia, himself a data journalist who had been part of the ICIJ's Panama Papers team, continued to investigate the very networks that had cost his mother her life.[14] Andrew and Paul Caruana Galizia, her other sons, became vocal advocates for press freedom and accountability. The family established a foundation in her name to support investigative journalism across Europe and to ensure that the work she had started would not die with her. Her final blog post, published on the afternoon she was killed, had ended with a line that became an epitaph for her work and a warning to anyone who sought to make the invisible visible. She wrote that there were crooks everywhere and that the situation was desperate.

The message her murder sent to other investigators was unmistakable. This was the cost of pulling back the ring. Financial invisibility protected not just wealth but the people who accumulated it, and those protections extended to violence when exposure threatened the system. The architecture of invisible money was not a passive structure. It was defended by people willing to kill.

The scale of hidden wealth is difficult to comprehend because the wealth itself is designed to resist measurement. Estimating how much money the world's wealthy have hidden from public view is like trying to count fish in a dark ocean. The tools of concealment are specifically engineered to prevent the kind of transparency that would make accounting possible.

Gabriel Zucman, an economist at the University of California, Berkeley, has produced the most rigorous estimates available. His research, published in "The Hidden Wealth of Nations," calculated that approximately $7.6 trillion in financial assets were held in tax havens as of 2017.[16] The figure represented roughly eight percent of the world's total financial wealth. Zucman estimated that the taxes lost on this hidden wealth amounted to more than $200 billion per year globally.[16]

These numbers were almost certainly conservative. Zucman's methodology relied on discrepancies between international financial data, comparing what countries reported as foreign-owned assets against what other countries reported as domestically held wealth. The gaps in these figures revealed the shadow economy of hidden money. But the methodology could only capture what the data could see. Money hidden through more sophisticated means likely escaped measurement entirely. Wealth concealed through layered shell companies, perpetual trust structures, cryptocurrency transactions, and informal transfer systems such as hawala networks left no statistical trace for economists to follow. The true figure could be significantly higher than $7.6 trillion.

Some researchers suggested it might be double that amount.[16]

The consequences of this hidden wealth extended far beyond lost tax revenue. Developing nations suffered disproportionately. African countries alone lost an estimated $50 billion per year to illicit financial flows, a figure that exceeded the total foreign aid those countries received.[17] The money flowed from the poorest nations on earth into the financial systems of the wealthiest, where it was laundered through real estate purchases in London and New York, through art auctions and luxury goods, through the very institutions that lectured the developing world about good governance and fiscal responsibility.

The Tax Justice Network, a research and advocacy organization based in the United Kingdom, developed the Financial Secrecy Index to rank countries by the degree of financial invisibility they provide.[18] The index evaluated jurisdictions on a combination of factors. The opacity of their corporate registration systems, the strength of their banking secrecy laws, the availability of trust structures, and the scale of financial activity flowing through their borders.

The results of the index were consistently surprising to people who associated financial secrecy with small Caribbean islands or obscure European principalities. The United States ranked as one of the world's most significant enablers of financial secrecy.[18] Delaware's corporate registration system allowed companies to be formed with minimal disclosure of beneficial ownership. Nevada offered similar advantages. South Dakota had emerged as a global destination for trust assets, attracting hundreds of billions of dollars from wealthy individuals worldwide with a combination of favorable laws and near-absolute confidentiality.

The United Kingdom ranked high on the index not because of secrecy within its own borders but because of the network of overseas territories and Crown dependencies it maintained.[18] The British Virgin Islands, the Cayman Islands, Jersey, Guernsey, and the Isle of Man all operated under the British Crown while offering levels of financial secrecy that London itself did not. This arrangement created a peculiar dynamic. The United Kingdom could present itself as a champion of financial transparency while its dependent territories functioned as some of the world's most prolific providers of financial invisibility.[7]

Switzerland, despite decades of international pressure and several rounds of reform, continued to rank among the most secretive financial jurisdictions on Earth. The numbered account had become less common, but the culture of secrecy that produced it remained deeply embedded in Swiss financial law and practice. Singapore had risen to rival Switzerland as a destination for hidden wealth, offering similar guarantees of confidentiality with the added attraction of proximity to the booming economies of Asia. Hong Kong served a similar function for Chinese capital seeking an exit from mainland restrictions. Luxembourg, nestled between France, Germany, and Belgium, hosted a financial sector whose assets dwarfed its tiny national economy. Each of these jurisdictions maintained systems that prioritized the privacy of wealth holders over the transparency demanded by democratic accountability.[18]

The paradox at the heart of this system was its ordinariness. Financial secrecy was not maintained by rogue states or lawless territories operating beyond the reach of international norms. It was maintained by jurisdictions that belonged to the democratic West, territories that participated in international organizations, countries that signed treaties and hosted summits about the importance of transparency and the rule of law. The architecture of invisible money was not a deviation from the global order. It was a feature built into the global order by design.

This was the Ring of Gyges operating at institutional scale. When Glaucon posed his thought experiment in Plato's Republic, he asked what a single person would do if granted the power to act without being seen. The offshore financial system answered the question for entire populations of wealthy individuals simultaneously. It provided a permanent, structural ring of invisibility to anyone who could afford the legal fees to access it. The answer to Glaucon's question was the same at the institutional level as it had been at the individual level. When people could act without being seen, many of them chose to stop paying taxes. They chose to hide assets from business partners, from spouses, from creditors, and from the governments of the countries where they lived. They chose self-interest over obligation, accumulation over contribution.

Not every offshore account holder was engaged in criminal activity. Legitimate reasons for holding assets in foreign jurisdictions existed. International businesses needed corporate structures that spanned multiple countries. Individuals living in unstable political environments had rational reasons to keep assets beyond the reach of unpredictable governments. The tools of financial invisibility served genuine purposes, just as they had from the very beginning.

But the scale of abuse dwarfed the legitimate use. The Panama Papers and the Pandora Papers together revealed more than 23 million documents and exposed hundreds of thousands of offshore entities.[11] The vast majority of these entities existed for a single purpose... to hide wealth from scrutiny. The system had been built for the innocent and seized by the powerful, following the same pattern that had repeated itself from bearer bonds to Bitcoin.

Gyges' kingdom invented money. The electrum coins stamped with the lion of Lydia were the first technology for separating value from identity, the first tool that allowed wealth to move through the world without a name attached to it. Centuries later, that technology has been refined beyond anything the Lydians could have imagined.

A teenager with a laptop can create a cryptocurrency wallet that is functionally untraceable. A law firm in a Caribbean island can establish a chain of shell companies that leads nowhere, each entity owning the next in an infinite regression of anonymity. A trust company in South Dakota can hold billions in assets while disclosing the beneficial owners to precisely no one. The tools have changed. The principle has not.

Money is a ring of invisibility. It has been one since the moment it was invented, since the first Lydian merchant accepted a coin from a stranger and did not ask for a name. Every innovation in the long history of finance has made that ring more powerful. Bearer bonds removed the need for identity in debt. Swiss bank accounts criminalized the disclosure of financial information. Shell companies created legal persons who could own wealth without revealing their creators. Offshore jurisdictions built entire economies around the service of concealment. Cryptocurrency promised to make the ring digital and permanent.

What began as a reasonable answer to a real problem became, in every instance, a gateway to abuse. The progression from invention to exploitation was not a failure of regulation or a gap in the law. It is a feature of human nature.

The estimated $7.6 trillion hidden in tax havens represents more than a fiscal problem.[16] It represents the answer to one of the oldest questions in philosophy. What do people become when no one is watching? The offshore financial system provides the data. When people can hide their wealth, many of them hide it. When people can avoid their obligations, many of them avoid them. When people can act without consequences, the logic is identical across cultures, across centuries, across every level of wealth and power.

Gyges found his ring in a cave beneath the earth. The modern ring hides in plain sight, in the corporate registries of Delaware and the trust companies of South Dakota, in the numbered accounts of Zurich and the blockchain transactions of a decentralized network that spans the globe. It is the most powerful ring ever created because it does not require magic or myth. It requires only money and the willingness to use the legal structures that the world's most respected democracies have built to protect it.

References

[1] Herodotus, *The Histories*, Book I. Herodotus provides the earliest surviving account of Lydian coinage and the story of Gyges. The connection between the invention of coins and the Gyges legend in the same Lydian culture is explored in Richard Seaford, *Money and the Early Greek Mind: Homer, Philosophy, Tragedy* (Cambridge: Cambridge University Press, 2004).

[2] The dating of the earliest Lydian electrum coins remains contested among numismatists. See David Schaps, *The Invention of Coinage and the Monetization of Ancient Greece* (Ann Arbor: University of Michigan Press, 2004), and the British Museum's catalog of early Lydian coinage. Most scholars place the first coins between 630 and 600 BC under the Mermnad dynasty.

[3] On the history of bearer bonds and their role in anonymous wealth transfer, see Nicholas Shaxson, *Treasure Islands: Uncovering the Damage of Offshore Banking and Tax Havens* (New York: Palgrave Macmillan, 2011).

[4] The Tax Equity and Fiscal Responsibility Act of 1982 (TEFRA), Pub. L. 97-248, effectively ended the issuance of bearer bonds in the United States by requiring registration of bonds for tax-deduction eligibility.

[5] The Swiss Banking Act of 1934 (Federal Act on Banks and Savings Banks) criminalized the disclosure of client information by Swiss bankers. On its origins in the context of Nazi persecution and its subsequent exploitation, see Sebastian Guex, "The Origins of the Swiss Banking Secrecy Law and Its Repercussions for Swiss Federal Policy," *Business History Review* 74, no. 2 (2000).

[6] On the history of shell companies, corporate layering, and beneficial ownership concealment, see the World Bank and United Nations Office on Drugs and Crime, *The Puppet Masters: How the Corrupt Use Legal Structures to Hide Stolen Assets and What to Do About It* (Washington,

DC: World Bank, 2011).

[7] On offshore jurisdictions and their role in global financial secrecy, including the British Virgin Islands, the Cayman Islands, and the network of British Crown dependencies, see Shaxson, *Treasure Islands*, and Oliver Bullough, *Moneyland: The Inside Story of the Crooks and Kleptocrats Who Rule the World*(New York: St. Martin's Press, 2019).

[8] Satoshi Nakamoto, "Bitcoin: A Peer-to-Peer Electronic Cash System," published October 31, 2008. Available at bitcoin.org/bitcoin.pdf. On the anonymity of Nakamoto and the ideological roots of Bitcoin, see Nathaniel Popper, *Digital Gold: Bitcoin and the Inside Story of the Misfits and Millionaires Trying to Reinvent Money* (New York: Harper, 2015).

[9] Bastian Obermayer and Frederik Obermaier, *The Panama Papers: Breaking the Story of How the Rich and Powerful Hide Their Money* (London: Oneworld Publications, 2016). The authors provide a detailed account of the initial contact from "John Doe," the encrypted communications, and the manifesto the source published after the leak.

[10] The total volume of the Panama Papers leak was 2.6 terabytes encompassing 11.5 million documents. See International Consortium of Investigative Journalists, "Giant Leak of Offshore Financial Records Exposes Global Array of Crime and Corruption," April 3, 2016, icij.org.

[11] International Consortium of Investigative Journalists, "The Panama Papers: Exposing the Rogue Offshore Finance Industry," April 3, 2016,

icij.org. The ICIJ coordinated the collaboration of nearly four hundred journalists from 107 media organizations in 80 countries. On the political fallout, including the resignations of Sigmundur Gunnlaugsson and the disqualification of Nawaz Sharif, see ICIJ's ongoing Panama Papers coverage archive.

[12] International Consortium of Investigative Journalists, "Pandora Papers," October 3, 2021, icij.org. The investigation drew on 11.9 million documents from 14 offshore service providers and revealed hidden wealth of world leaders including King Abdullah II of Jordan and Czech Prime Minister Andrej Babis.

[13] On the United States as a destination for hidden wealth, particularly South Dakota trust companies and Delaware corporate formation, see Jesse Drucker and Eric Lipton, "How the Biggest U.S. Tax Haven Helps the Richest Get Richer," *The New York Times*, August 14, 2021, and the ICIJ Pandora Papers reporting on American secrecy jurisdictions.

[14] On Daphne Caruana Galizia's investigative journalism in Malta, her blog Running Commentary, and her Panama Papers-related reporting on Konrad Mizzi and Keith Schembri, see Manuel Delia, *Murder on the Malta Express: Who Killed Daphne Caruana Galizia?* (London: Midsea Books, 2019), and the Daphne Caruana Galizia Foundation archives at daphne.foundation.

[15] On the assassination of Daphne Caruana Galizia, the subsequent investigation, the arrests of Vince Muscat, Alfred and George Degiorgio,

and Yorgen Fenech, and the 2021 public inquiry findings, see the Council of Europe report, "Justice for Daphne Caruana Galizia," and the Maltese Public Inquiry Board of Inquiry report, July 29, 2021. On the SLAPP suits and intimidation campaign, see Reporters Without Borders, "Malta: RSF Files Complaint over Daphne Caruana Galizia Murder," 2017.

[16] Gabriel Zucman, *The Hidden Wealth of Nations: The Scourge of Tax Havens* (Chicago: University of Chicago Press, 2015). Zucman estimated approximately $7.6 trillion in financial assets held in tax havens as of 2014, with global tax revenue losses exceeding $200 billion per year. Updated estimates appear in Zucman's subsequent research with Annette Alstadsaeter and Niels Johannesen.

[17] On illicit financial flows from developing nations, see Global Financial Integrity, *Illicit Financial Flows from Developing Countries* (Washington, DC: GFI, 2015), and the African Union/UN Economic Commission for Africa High Level Panel on Illicit Financial Flows from Africa (the Mbeki Report), 2015.

[18] Tax Justice Network, *Financial Secrecy Index*, published biennially at fsi.taxjustice.net. The index ranks jurisdictions by a combination of secrecy scoring and the scale of financial services activity, consistently placing the United States, the United Kingdom (via its overseas territories), and Switzerland among the most significant enablers of global financial secrecy.

Chapter Eight - Shielded

On March 20, 1602, a group of Dutch merchants did something that had never been done before. They pooled their capital into a single enterprise called the Vereenigde Oostindische Compagnie, the United East India Company, and they sold shares of ownership to anyone willing to buy them. The investors could profit if the company succeeded. They could lose their investment if it failed. But they could not lose anything more than that. Their homes, their other assets, their personal fortunes beyond the shares themselves were protected by a wall that no creditor could breach.[1] This was the invention of limited liability. It was, by any reasonable measure, one of the most consequential innovations in the history of human organization. It unlocked the capacity for large-scale economic cooperation. It allowed thousands of strangers to fund ventures they would never personally oversee, from spice routes across the Indian Ocean to railroads across the American continent. The logic was sound and the purpose was legitimate. If people had to risk everything they owned every time they backed a commercial venture, very few people would ever back anything. Limited liability made modern capitalism possible.

But it also did something else. It created a structure in which the people who made decisions were separated from the people who suffered the consequences of those decisions. The investors were shielded from personal ruin. The managers were shielded from personal blame. Over the centuries that followed, the concept evolved and hardened into legal doctrine. By the nineteenth century, courts in

England and the United States had established the principle of corporate personhood, the idea that a corporation is a legal person in its own right. This legal person could sign contracts, own property, incur debts, and even commit crimes. It could be taken to court and forced to pay damages. But this legal person had no body to imprison and no conscience to trouble. It could not be shamed at a dinner party or confronted by a grieving mother in a courtroom hallway. It was an abstraction with the rights of a human being and the accountability of a ghost. When the corporation caused harm, the corporation was punished. The people inside the corporation, the ones who actually made the choices that produced the harm, stood behind it like soldiers behind a fortress wall.

Consider the implications through the lens of Glaucon's question. The corporation did not give any single person the power of invisibility. Instead, it distributed that power across hundreds or thousands of people simultaneously. Every individual within the structure could point to someone else. The executive could point to the board. The board could point to the shareholders. The manager could point to the executive who set the policy. The employee could point to the manager who gave the order. Responsibility flowed in circles, like water in a drain, until it disappeared entirely.

This chapter examines four cases in which that disappearing act produced catastrophic results. The cases span different industries, different countries, and different decades. In each case, the people involved were not cartoon villains. They were professionals operating within systems that made it easy to cause enormous harm without ever feeling personally responsible for any of it. The ring they wore was not forged in fire or found in a cave. It was structural. It was designed, it was legal, and it was devastatingly effective.

In 1995, the Food and Drug Administration approved a new painkiller called OxyContin. The drug was manufactured by Purdue Pharma, a private pharmaceutical company headquartered in Stamford, Connecticut. Its active ingredient was oxycodone, a powerful opioid that had been used in medicine for decades. What made OxyContin different was its delivery system. The drug was formulated with a time-release coating that delivered the opioid gradually over twelve hours, rather than all at once. Purdue's marketing centered on a single revolutionary claim. Because of this controlled-release mechanism, OxyContin carried a lower risk of addiction than other opioids.[2]

The claim was not true. Purdue knew it was not true. Internal documents later revealed through litigation showed that the company had tracked reports of abuse, addiction, and overdose deaths from the earliest days of OxyContin's release and had systematically concealed this information from regulators and the public.[2] Purdue assembled a sales force of more than a thousand representatives and deployed them across the country with a singular mission. They were trained to tell doctors that the risk of addiction from OxyContin was less than one percent, a figure that had no credible scientific basis. Their instructions included recommending higher doses for patients who reported breakthrough pain, even though higher doses meant greater dependency. Each representative carried scripts and talking points that minimized every warning sign. When doctors expressed hesitation about prescribing a powerful opioid for chronic conditions like back pain or arthritis, the sales representatives had answers prepared for every objection.[2] The entire operation was built on the premise that hesitation was a sales problem, not a medical one.

The family behind Purdue Pharma was the Sacklers. Arthur, Mortimer, and Raymond Sackler had built the company over decades, and their descendants controlled it absolutely. The Sackler family did not merely own Purdue in the way that shareholders own a public company. They ran it. They sat on its board. They reviewed its marketing strategies. They approved of its messaging. They directed the campaign that positioned OxyContin as a safe, broadly applicable treatment for chronic pain.[2] Internal emails revealed during litigation showed members of the Sackler family discussing sales targets, market expansion, and strategies for dealing with the growing evidence of addiction. One email chain included discussions of overdose deaths framed not as a human tragedy but as a public relations problem that needed to be managed so it would not reduce sales.[2]

The scale of what followed is almost impossible to comprehend in human terms. Between 1999 and 2020, more than 500,000 people in the United States died of opioid overdoses.[3] That figure, five hundred thousand dead, is larger than the population of most American cities. It exceeds the number of Americans killed in World War II. It includes deaths from OxyContin itself, from generic oxycodone that flooded the market after Purdue's success, from heroin that people turned to when their prescriptions ran out, and from fentanyl that eventually replaced heroin as the cheapest way to feed an addiction that often began with a legitimate prescription. Not every one of those deaths can be traced directly to Purdue Pharma. But Purdue Pharma lit the match. The company created the market for mass opioid prescribing. It funded the research that claimed chronic pain was undertreated. It built the sales force that convinced doctors to prescribe more freely. It suppressed the evidence that the drug was destroying lives. Each of those 500,000 deaths was a person with a name, a family, a life that was interrupted and then ended by a chemical dependency that someone in a Stamford, Connecticut office building had decided to profit from.

Throughout all of this, the Sackler name appeared on museum wings at the Metropolitan Museum of Art, the Louvre, and the Smithsonian. It appeared on buildings at Harvard, Yale, Oxford, and dozens of other universities. It did not appear on OxyContin bottles. For the general public, there was no connection between the family whose philanthropy decorated the world's most prestigious institutions and the company whose product was fueling an epidemic of addiction and death. The corporate structure stood between the Sacklers and their consequences like a castle wall.[2]

When the reckoning finally came, it came slowly and incompletely. Purdue Pharma filed for bankruptcy in 2019. The Sackler family agreed to pay approximately six and a half billion dollars toward settlements with states, cities, and tribal nations. In exchange, the original bankruptcy plan granted the Sacklers personal immunity from all future opioid-related lawsuits. A family that had extracted an estimated thirteen billion dollars from the company would pay less than half of that back, and then no one could ever sue them again.[2] The Supreme Court partially reversed this arrangement in 2024 in Harrington v. Purdue Pharma, ruling that the bankruptcy court did not have the authority to grant such sweeping immunity to parties who had not themselves filed for bankruptcy.[4] But the ruling was narrow. It addressed the legal mechanism, not the underlying injustice. No member of the Sackler family went to prison. No member of the Sackler family was criminally charged. The people who made the decisions that contributed to the deaths of hundreds of thousands of Americans remained wealthy, free, and largely anonymous to the general public.

The ring worked precisely as designed. Individual human beings made choices that caused enormous suffering. The corporate structure absorbed the blame, paid the fines, and shielded the individuals from meaningful personal consequences. Consider the stark asymmetry of this outcome. A street-level drug dealer who sells opioids to a person who later overdoses can be charged with manslaughter. A pharmaceutical executive who approves a marketing campaign that leads to hundreds of thousands of overdose deaths pays a settlement and keeps his house in Connecticut. The difference is not in the harm caused. The difference is in the structure that stands between the decision-maker and the consequence. If Richard Sackler had stood on a street corner handing out addictive pills while lying about their dangers, he would have been arrested within the hour. When he did the equivalent thing through a corporation, it took two decades and an act of the Supreme Court to produce even partial accountability.

On October 29, 2018, Lion Air Flight 610 departed Jakarta, Indonesia, bound for Pangkal Pinang. The flight was expected to last about an hour. Thirteen minutes after takeoff, the Boeing 737 MAX 8 aircraft plunged into the Java Sea at a speed that left no survivors and almost no recognizable wreckage. All 189 people on board were killed.[5] Less than five months later, on March 10, 2019, Ethiopian Airlines Flight 302 departed Addis Ababa bound for Nairobi. Six minutes after takeoff, the Boeing 737 MAX 8 crashed into a field southeast of the city with such force that it created a crater ten meters deep. All 157 people on board were killed.[5] The two crashes killed 346 people in total across 35 different nationalities. They shared the same cause.

Both aircraft were brought down by a software system called MCAS, the Maneuvering Characteristics Augmentation System. Boeing had designed the 737 MAX as an update to its best-selling 737 line, fitting it with larger, more fuel-efficient engines. The new engines changed the aerodynamic characteristics of the plane in a way that could cause the nose to pitch upward under certain conditions. Rather than redesigning the aircraft to fix this problem, Boeing created MCAS as a software patch. The system would automatically push the nose of the aircraft down if sensors detected that the angle of attack was too steep.[6]

The fundamental flaw was that MCAS relied on data from a single angle-of-attack sensor. Modern aviation safety philosophy demands redundancy for critical systems. If one sensor fails, another sensor provides backup data. MCAS had no such backup. If that single sensor malfunctioned, and sensors do malfunction, MCAS would push the nose down even when the plane was flying normally. The pilots would pull back on the controls. MCAS would push the nose down again. The pilots would pull back again. MCAS would push harder. In the cockpit of Lion Air Flight 610, Captain Bhavye Suneja and First Officer Harvino fought against the system for thirteen minutes as the aircraft pitched violently up and down over the Java Sea. They did not know what was fighting them. They had never been told about MCAS. In both crashes, this deadly tug-of-war between human beings and software ended with the software winning and the airplane diving into the ground at catastrophic speed.[5],[6]

Boeing knew about the risks. Internal documents released by the U.S. House Committee on Transportation and Infrastructure revealed a pattern of deliberate concealment that rivaled anything at Purdue Pharma.[6] Engineers within Boeing had raised concerns about MCAS during the development process. They warned that the system's reliance on a single sensor created a dangerous single point of failure. They recommended changes that would have added redundancy and improved safety. Management overruled them. The changes would have added cost, delayed the certification timeline, and potentially required pilots to undergo expensive simulator training before flying the new aircraft.[6]

This last point was critical. Boeing was locked in fierce competition with Airbus, whose A320neo was outselling the 737 MAX. One of Boeing's key selling points to airlines was that pilots already certified on older 737 models could fly the MAX without simulator training. If MCAS required simulator time, that selling point would collapse. Southwest Airlines alone had made its purchase contingent on avoiding the simulator requirement.[6] So Boeing made MCAS training optional. The company did not adequately disclose the system's existence or its potential behavior to pilots or airlines. Most of the pilots who flew the 737 MAX did not know MCAS existed until after the first crash.

The internal culture at Boeing had shifted dramatically in the years before the MAX disasters. In 1997, Boeing merged with McDonnell Douglas, a company known for prioritizing financial metrics over engineering culture. Former Boeing engineers described the merger as a hostile takeover of Boeing's engineering values by McDonnell Douglas's business values.[6] Internal communications released to Congress captured the resulting atmosphere with devastating clarity. One Boeing employee wrote that the 737 MAX was "designed by clowns, who are in turn supervised by monkeys."[7] Another described the culture as one in which raising safety concerns was career suicide. The message was clear... meet the deadline, beat the competition, and keep the costs down. Safety was important, but not as important as schedule.

The regulatory failure compounded the corporate one. The Federal Aviation Administration had delegated significant portions of the 737 MAX's safety certification to Boeing itself through a program called Organization Designation Authorization.[6] Boeing employees were simultaneously working for the company and serving as the company's safety inspectors. They were the builder and the building inspector at the same time. When those employees flagged problems, their concerns were filtered through Boeing's management chain before reaching the FAA. The fox was not merely guarding the henhouse. The fox was conducting the henhouse inspection and writing the henhouse safety report.

After both crashes, the 737 MAX was grounded worldwide for nearly two years. Boeing paid 2.5 billion dollars in a deferred prosecution agreement with the Department of Justice.[6] The agreement included a criminal fine of $243.6 million, compensation of $1.77 billion to the airlines that had purchased the MAX, and $500 million for a crash-victim beneficiaries fund. The families of the 346 dead received their portions of that fund. No individual Boeing executive was imprisoned. The engineers who raised concerns and were overruled faced no consequences because they had tried to do the right thing. The managers who overruled them faced no consequences because the corporate structure diffused their decisions into collective institutional action. The 346 people who died, including families traveling together, newlyweds, aid workers, students returning to school, had no voice in any of it.

The structure mirrored what came before. Individual people made specific decisions. Rely on a single sensor, skip simulator training, conceal MCAS from pilots and prioritize schedule over safety. Each of those decisions had a name attached to it somewhere inside Boeing's organizational chart. But the corporate structure transformed individual choices into institutional outcomes. By the time the decisions reached the outside world, they were Boeing's decisions, not any single person's decisions. The ring of the corporation made every decision-maker invisible.

This story is a personal favorite of mine. The level of deception and engineering could have spawned a Hollywood masterpiece. The lengths people will go to cheat are astounding to me. In the German city of Wolfsburg, Lower Saxony, the Volkswagen Group operates one of the largest automobile manufacturing complexes on Earth. The factory covers more than six square kilometers. It has its own power plant, its own rail network, and its own zip code. It produces roughly 800,000 vehicles per year.[8] Volkswagen has been Germany's largest employer and its most iconic brand for generations. When the company speaks, the German government listens. When the company stumbles, the entire German economy feels it.

Between 2009 and 2015, Volkswagen installed software in

approximately 11 million diesel vehicles worldwide that constituted one of the most elaborate consumer frauds in automotive history.[9] The software was designed to detect when a car was being subjected to an emissions test. During testing, the software activated the vehicle's full suite of pollution controls, ensuring the car met regulatory standards for nitrogen oxide emissions. During normal driving, the software deactivated or reduced those controls. The result was that Volkswagen's "clean diesel" vehicles emitted up to 40 times the legal limit of nitrogen oxides during everyday use.[9]

The deception was elegant in its simplicity. Emissions testing takes place on a dynamometer, a machine that simulates driving conditions while the car remains stationary. The software recognized the signature pattern of a dynamometer test by monitoring factors like wheel speed, engine operation, and steering wheel position. When it detected that the car was being tested, it switched into a clean mode. When it detected normal driving conditions, it switched back to a dirty mode that prioritized engine performance over emissions compliance.[8]

No government regulator discovered the fraud. The Environmental Protection Agency did not discover it. The California Air Resources Board did not discover it. The German Federal Motor Transport Authority did not discover it. For six years, eleven million vehicles rolled off assembly lines carrying software designed to deceive the agencies responsible for protecting public health, and not a single inspector caught it. The deception was uncovered by a small team of researchers at West Virginia University who were conducting an independent study of diesel emissions in 2014.[8] The researchers had expected Volkswagen's diesel cars to perform well, since the company's marketing heavily emphasized their clean credentials. Instead, they found enormous discrepancies between laboratory results and real-world driving data. The numbers were not slightly off. They were off by a factor of forty. The researchers published their findings. The EPA investigated. Volkswagen initially denied everything, insisting that the discrepancies must be the result of testing errors or unusual driving conditions. Then the company partially admitted to the use of what it called "defeat devices." Then the full scope of the scheme became public, and the largest automaker in Europe found itself at the center of the largest automotive scandal in history.

The internal culture at Volkswagen helped explain how such a massive fraud could persist for six years without anyone blowing the whistle. Investigators and journalists described VW's corporate culture as rigidly hierarchical, with a management style rooted in intimidation.[8] Ferdinand Piech, who served as CEO and later as chairman of the supervisory board, was legendary for his demanding temperament and his intolerance of failure. The culture he built rewarded results and punished dissent. Engineers who questioned directives risked their careers. Employees described a "fear culture" in which the implicit message from leadership was to achieve the stated targets by whatever means necessary, without burdening management with the details of how it was done.[8]

When the fraud was exposed, the engineers who had coded the defeat device software testified that they were following orders.[8] They had been told to make the cars pass emissions tests while also delivering the engine performance that marketing had promised. The two goals were physically incompatible. The only way to achieve both was to cheat.

The engineers said they raised concerns. They said they were told to find a solution. They found one. It was illegal, but it worked. The managers who issued the directives said they did not know the specific technical details of how their targets were being met. They set goals. Their subordinates achieved those goals. The method was not their department.

This was the corporate ring in its purest form. At the bottom of the hierarchy, an engineer sat at a workstation and wrote lines of code that would detect a dynamometer test and switch the engine into compliance mode. That engineer could say with complete sincerity that he was doing what he was told to do. His manager could say with equal sincerity that he did not specify the technical method. The division head could say that he set performance targets without dictating implementation. The CEO could say that he relied on his team to achieve results within the bounds of the law. The people who wrote the code could point upward. The people who gave the orders could point downward. Responsibility dissolved at every level of the hierarchy like sugar in hot water. CEO Martin Winterkorn was eventually charged with fraud, but his trial was repeatedly delayed, and proceedings were ultimately suspended due to his declining health.[9] A handful of engineers were convicted. Volkswagen paid more than 30 billion dollars in fines, settlements, and vehicle buybacks across multiple countries.[9] The company survived. The brand survived. The executives who presided over the fraud largely retired with their fortunes intact.

The numbers were staggering. But numbers, even staggering ones, can feel abstract when they appear in a settlement agreement rather than on a death certificate. Nitrogen oxides are not an abstract pollutant.

They cause respiratory disease, cardiovascular damage, and premature death. Researchers estimated that the excess emissions from Volkswagen's defeat devices contributed to approximately 1,200 premature deaths in Europe alone over the period the fraud was active.[8] Those deaths were statistical, not dramatic. Nobody collapsed in the street clutching a Volkswagen exhaust pipe. The deaths showed up as marginal increases in hospital admissions for asthma, in elevated rates of chronic bronchitis, in shortened lifespans for people who already had compromised respiratory systems. The people who died never knew that a software engineer in Wolfsburg had contributed to their deaths. The software engineer in Wolfsburg never knew that any specific person had died because of what he wrote. The corporate structure ensured that cause and consequence never met.

The headquarters of Wells Fargo occupies a cluster of towers in downtown San Francisco. The company is the fourth-largest bank in the United States by assets. It serves approximately one in three American households. Its stagecoach logo is one of the most recognized symbols in American finance. For more than a decade, that logo concealed a fraud so pervasive that it touched millions of customers who never knew they had been victimized.[10]

Between 2002 and 2016, over the course of fourteen years, Wells Fargo employees created approximately 3.5 million unauthorized bank accounts, credit card accounts, and other financial products in customers' names without their knowledge or consent.[11] Customers who walked into a Wells Fargo branch to open a single checking account would later discover that they also had a savings account they had never requested, a credit card they had never applied for, and an online banking profile they had never activated. Some customers were charged fees on these phantom accounts. Others had their credit scores damaged. Others simply never noticed, which was the point.

The fraud was driven by a sales culture that had been deliberately designed from the top of the organization. Wells Fargo's retail banking division operated under a program called "Going for Gr-Eight," which encouraged employees to sell eight financial products to every customer.[10] Branch managers received aggressive sales quotas. Employees who met their quotas received bonuses. Employees who failed to meet their quotas were placed on performance improvement plans, publicly humiliated in team meetings, and eventually terminated.[11] The pressure was relentless and it was intentional. The message from corporate leadership was that cross-selling was the engine of Wells Fargo's growth strategy. Every customer interaction was an opportunity to sell more products. Every product sold was a data point that flowed up the chain to Wall Street analysts, who rewarded Wells Fargo's stock price for its industry-leading cross-sell ratio.

The employees who created the fake accounts were not criminal masterminds. They were bank tellers, customer service representatives, and personal bankers who earned between $12 and $25 per hour.[11] Most of them were people who simply needed their jobs. When the quota system made it impossible to sell eight products to customers who only wanted one or two, the employees faced a choice. Fail to meet their quotas, endure the consequences, and eventually lose their jobs. Or create unauthorized accounts to pad their numbers and keep their paychecks. Thousands of them chose the second option. The fraud was so widespread that it became normalized within the culture. New employees learned the techniques from veteran employees who treated unauthorized account creation as an open secret that everybody did.

Whistleblowers tried to sound the alarm. Internal reports documenting the unauthorized accounts reached senior management years before the scandal became public.[10] Employees who reported the fraud through internal channels were sometimes fired or demoted. One employee in a Riverside, California branch reported the unauthorized account creation to a manager and was terminated within weeks.[11] The company's ethics hotline received thousands of complaints over the years. The volume alone should have triggered an institutional crisis. Instead, senior leaders acknowledged the complaints in private meetings and continued the sales pressure in public earnings calls. The cross-sell ratio was too important to Wall Street. It was the number that analysts cited when recommending Wells Fargo stock. It was the metric that justified the premium valuation. The fraud was not a bug in the system. It was the system's primary output.

When the scandal finally broke in 2016, Wells Fargo fired approximately 5,300 low-level employees for creating unauthorized accounts.[10] These were the tellers, the customer service representatives, and the personal bankers who had implemented the fraud at the branch level. CEO John Stumpf testified before the Senate Banking Committee and claimed that the unauthorized accounts were the work of rogue employees who had violated the company's values.[10] He described the problem as a failure of individual ethics, not a failure of corporate design.

The evidence contradicted him completely. The sales quotas had been set by senior management. The cross-sell strategy had been designed by senior management. The pressure system that made fraud inevitable had been built by senior management. The internal complaints that documented the fraud had been received and ignored by senior management.[11] The entire apparatus was a machine designed to produce exactly the outcome it produced, and the people who designed the machine pointed at the people who were ground up inside it.

Stumpf eventually resigned. He left with a retirement package initially valued at approximately $134 million, though regulators later clawed back a portion of it.[10] Carrie Tolstedt, the senior vice president who had overseen the retail banking division during the worst period of the fraud, initially left with a $125 million payout.[11] She was later barred from the banking industry by regulators and faced criminal charges. But the vast majority of senior executives who had benefited from the cross-sell strategy faced no significant consequences. The 5,300 low-level employees who were fired had their careers destroyed. Many struggled to find work in banking again because their termination records showed they had been fired for creating fraudulent accounts.

The cruelty of this outcome deserves emphasis. The system was designed from the top. The pressure was applied from the top. The quotas were set from the top. The warnings were ignored at the top. The profits flowed to the top. When the fraud was exposed, the punishment fell on the bottom. A bank teller making $14 an hour lost her career for doing what the institution demanded of her. A CEO making $19 million a year resigned with a golden parachute for building the institution that demanded it. The people who had the least power to resist the system bore the greatest consequences for participating in it. The people who built the system walked away with their fortunes. This was not a failure of the corporate shield. This was the corporate shield working exactly as intended. The ring did not merely protect the powerful from external consequences. It redirected those consequences downward onto the powerless.

Pull back from the details of these four cases and a shared architecture becomes visible with the clarity of a geological formation. The specifics differ in every possible way. Pharmaceuticals in one case, aircraft in another, automobiles in a third, banking in a fourth. The industries are unrelated. The products are unrelated. The countries and regulatory frameworks overlap only partially. One case involves a private family-owned company while another involves a publicly traded multinational. One case produced half a million deaths over two decades while another produced millions of cases of financial fraud that killed no one directly. Yet the underlying structure is identical in every case.

First, individual human beings made decisions they would not have made if their names had been personally and publicly attached to the outcomes. The Sackler family members who approved the marketing of OxyContin as a low-addiction drug would not have stood in front of a crowd of chronic pain patients and made that claim to their faces, knowing it was false. The Boeing managers who decided to skip simulator training for MCAS would not have boarded one of their own planes with their own families if they had been forced to confront the full implications of that decision. The Volkswagen executives who demanded emissions targets that could only be met through fraud would not have personally installed defeat devices in cars if it meant their names would appear on the software. The Wells Fargo leaders who built the cross-sell machine would not have personally opened fake credit cards in their neighbors' names. Anonymity changed behavior. The structure provided anonymity.

Second, the corporate hierarchy distributed responsibility so widely that no single person bore it. Purdue Pharma employed thousands of people. Boeing employed more than 150,000. Volkswagen employed more than 600,000. Wells Fargo employed more than 250,000. Within these vast organizations, every decision passed through multiple layers of review, approval, delegation, and implementation. The person who conceived a strategy was not the person who approved it. The person who approved it was not the person who implemented it. The person who implemented it was not the person who enforced it. At every link in the chain, the individual could point to the link above or below and say that the real decision was made somewhere else. The result was a system in which enormous harm was produced and no individual felt responsible for producing it.

Third, when the harm was finally revealed, the organization absorbed the punishment while the individuals largely walked away. Purdue Pharma filed for bankruptcy. Boeing paid a deferred prosecution agreement. Volkswagen paid record fines. Wells Fargo paid billions in penalties. These were not trivial consequences. They represented real financial losses for the organizations involved. But organizations do not suffer. Organizations do not grieve. Organizations do not lose sleep, lose their freedom or lose their sense of themselves as decent people. The financial penalties were distributed across shareholders, employees, and future customers. The individuals who made the decisions that produced the harm experienced these consequences as abstractions, as numbers on a balance sheet, as headlines in a newspaper, not as personal accountability for personal choices.

Finally, the people who suffered the most had the least visibility into what was happening. The opioid patients who became addicted to OxyContin did not know that the drug's safety profile had been fabricated. The passengers who boarded Lion Air Flight 610 and Ethiopian Airlines Flight 302 did not know that a hidden software system could override their pilots. The owners of Volkswagen diesel cars did not know that their vehicles were poisoning the air at 40 times the legal limit. The Wells Fargo customers whose names appeared on unauthorized accounts did not know those accounts existed. In each case, the victims were kept in the dark by the same structure that shielded the decision-makers from consequences. The invisibility ran in both directions. The corporation made the executives invisible to the public, and it made the harm invisible to the victims until it was too late.

Glaucon's question echoed through each of these boardrooms, though no one inside them heard it. The corporation provides a collective version of the ring's power. It does not make any single person invisible the way the shepherd became invisible. Instead, it creates an environment in which every person can believe that they are not the one who is truly responsible. The executive believes the board set the direction. The board believes the executive handled the details. The manager believes the executive set the policy. The employee believes the manager gave the order. Everyone is visible to someone, but no one is visible as the author of the harm.

The corporation does not need to corrupt the people inside it. This distinction matters. The Sacklers were not transformed into monsters by

Purdue Pharma. The Boeing managers were not twisted into villains by the company's culture. The Volkswagen engineers were not born with a desire to cheat emissions tests. The Wells Fargo tellers were not natural-born fraudsters. In most cases, the people involved would have been unremarkable neighbors, competent professionals, loving parents. The corporation did not change who they were. It changed what they could do without feeling the weight of it. It removed the link between decision and consequence. It put distance between the hand that signed the memo and the life that ended because of what the memo authorized.

That removal was enough. A Sackler family member could read about opioid deaths in the newspaper and feel genuine sadness without ever connecting that sadness to the marketing strategy she had approved in a boardroom six months earlier. A Boeing manager could attend the memorial service for the Lion Air crash victims and feel genuine grief without connecting that grief to the meeting where he had voted to skip simulator training. The distance between the decision and the death was filled with layers of corporate process that absorbed the moral weight the way insulation absorbs sound. By the time the consequence arrived, it was muffled beyond recognition.

The ring does not need to make you evil. It only needs to make you comfortable enough to look the other way. It needs to create just enough ambiguity that you can tell yourself the decision was not really yours, the harm was not really foreseeable, the responsibility really belongs to someone else. The corporation provides that ambiguity at industrial scale. It manufactures moral distance the way it manufactures products, efficiently and in enormous quantities.

There is a temptation at this point to frame the argument as an attack on corporations themselves, to suggest that the solution is the abolition of limited liability or the dismantling of large organizations. That temptation should be resisted. The limited liability corporation is one of the most effective tools for organizing human cooperation ever invented. It has produced extraordinary prosperity and innovation. The problem is not the tool. The problem is the invisibility the tool provides. A hammer can build a house or crack a skull. The question is not whether we should have hammers. The question is whether we have built adequate systems for holding people accountable when they use the hammer to cause harm.

In each of the four cases examined in this chapter, the accountability systems failed. Regulators were captured, underfunded, or deliberately circumvented. The FDA trusted Purdue's own studies on addiction risk. The FAA let Boeing certify its own aircraft. European regulators accepted Volkswagen's emissions data at face value. Banking regulators received warnings about Wells Fargo for years and took no meaningful action until journalists forced the issue into public view. Whistleblowers were punished rather than protected. Legal consequences fell on the organization rather than on the individuals who made the decisions. Criminal prosecution of executives was rare, slow, and often unsuccessful. The structural invisibility of the corporation was reinforced at every turn by a legal and regulatory environment that treated corporate misconduct as a civil matter rather than a criminal one. Fines were calculated as a cost of business. Settlement agreements were negotiated behind closed doors. Deferred prosecution agreements allowed corporations to admit wrongdoing without any individual admitting to anything at all.

The result is a world in which the consequences for corporate harm are almost always financial, almost never personal, and almost never proportional to the damage caused. A person can contribute to the deaths of hundreds of thousands of people and retire with a fortune. An executive can make decisions that crash airplanes and walk away without seeing the inside of a courtroom. A bank can defraud eleven million customers and have the punishment fall on the employees who were forced to do the defrauding. A manufacturer can poison the air in cities across Europe and pay a fine that amounts to a cost of doing business.

That is what makes this version of the ring so much more dangerous than the original. It does not feel like invisibility at all. It feels like a normal day at the office.

References

[1] On the founding of the Dutch East India Company (Vereenigde Oostindische Compagnie, or VOC) on March 20, 1602, and the invention of limited liability and tradeable shares, see Lodewijk Petram, *The World's First Stock Exchange*, trans. Lynne Richards (New York: Columbia University Press, 2014), and Stephen Bown, *Merchant Kings: When Companies Ruled the World, 1600-1900* (New York: Thomas Dunne Books, 2009).

[2] On Purdue Pharma, the Sackler family, OxyContin's marketing, internal documents revealing concealment of addiction data, and the bankruptcy settlement, see Patrick Radden Keefe, *Empire of Pain: The Secret History of the Sackler Dynasty* (New York: Doubleday, 2021), and

Barry Meier, *Pain Killer: An Empire of Deceit and the Origin of America's Opioid Epidemic*, rev. ed. (New York: Random House, 2018). Internal Sackler family emails and litigation documents were made public through proceedings in *In re Purdue Pharma L.P.*, U.S. Bankruptcy Court, Southern District of New York.

[3] Centers for Disease Control and Prevention, "Understanding the Opioid Overdose Epidemic," National Center for Injury Prevention and Control, updated 2023. The CDC reported more than 500,000 opioid-involved overdose deaths in the United States between 1999 and 2020.

[4] *Harrington v. Purdue Pharma L.P.*, 603 U.S. (2024). The Supreme Court ruled that the bankruptcy court lacked authority to grant nonconsensual third-party releases shielding the Sackler family from future opioid litigation.

[5] On the Lion Air Flight 610 crash (October 29, 2018, 189 killed) and Ethiopian Airlines Flight 302 crash (March 10, 2019, 157 killed), see Komite Nasional Keselamatan Transportasi (KNKT), *Aircraft Accident Investigation Report: PT. Lion Mentari Airlines Boeing 737-8 (MAX)*, October 2019, and Ethiopian Accident Investigation Bureau, *Aircraft Accident Investigation Preliminary Report*, March 2019.

[6] On the Boeing 737 MAX, the MCAS system, internal decision-making, the McDonnell Douglas merger's cultural impact, FAA delegation of certification authority, and the $2.5 billion deferred prosecution agreement, see U.S. House Committee on Transportation and

Infrastructure, *Final Committee Report: The Design, Development, and Certification of the Boeing 737 MAX*, September 2020, and Peter Robison, *Flying Blind: The 737 MAX Tragedy and the Fall of Boeing* (New York: Doubleday, 2021).

[7] The "designed by clowns, who are in turn supervised by monkeys" quote appeared in internal Boeing communications released to the U.S. House Committee on Transportation and Infrastructure in January 2020. See House Committee on Transportation and Infrastructure, "Boeing Withheld Documents from Investigators," press release, January 9, 2020.

[8] On the Volkswagen Wolfsburg factory, the technical mechanism of the defeat device software, the West Virginia University research that uncovered the fraud, the internal "fear culture" at VW, and the estimated 1,200 premature deaths from excess nitrogen oxide emissions, see Jack Ewing, *Faster, Higher, Farther: How One of the World's Largest Automakers Committed a Massive and Stunning Fraud* (New York: W.W. Norton, 2017), and Guillaume P. Chossiere et al., "Public Health Impacts of Excess NOx Emissions from Volkswagen Diesel Passenger Vehicles in Germany," *Environmental Research Letters* 12, no. 3 (2017).

[9] On the Volkswagen Dieselgate legal proceedings, including the scope of the fraud (11 million vehicles), CEO Martin Winterkorn's indictment and suspended trial, and the more than $30 billion in total fines and settlements, see the U.S. Environmental Protection Agency enforcement case records and the German Federal Court of Justice (Bundesgerichtshof) proceedings.

[10] On the Wells Fargo unauthorized accounts scandal, the "Going for Gr-Eight" cross-selling program, whistleblower retaliation, CEO John Stumpf's testimony before the Senate Banking Committee and subsequent resignation, see U.S. Senate Committee on Banking, Housing, and Urban Affairs, *An Examination of Wells Fargo's Unauthorized Accounts and the Regulatory Response*, hearing transcript, September 20, 2016, and Bethany McLean, "How Wells Fargo's Cutthroat Corporate Culture Allegedly Drove Bankers to Fraud," *Vanity Fair*, May 31, 2017.

[11] On the scale of unauthorized account creation (approximately 3.5 million accounts), employee working conditions and wage levels, the firing of 5,300 low-level employees, retaliation against the Riverside whistleblower, and the departure packages of senior executives including Carrie Tolstedt, see Consumer Financial Protection Bureau, Consent Order in the Matter of Wells Fargo Bank, N.A., File No. 2016-CFPB-0015, September 8, 2016, and Office of the Comptroller of the Currency enforcement actions against Wells Fargo, 2018-2020.

Chapter Nine - The Algorithm Behind the Curtain

In January of 2012, a team of researchers at Facebook quietly altered the News Feeds of 689,003 users. The experiment was simple in design and vast in scope. One group of unwitting participants saw fewer positive posts from their friends and family. Another group saw fewer negative posts. The researchers then measured whether this invisible manipulation changed the way those people expressed themselves online. The answer was yes. Users who saw fewer positive posts began writing more negative posts themselves. Users who saw fewer negative posts began writing more positively. The emotional weather of nearly 700,000 people shifted according to a dial that someone at Facebook had turned without telling them.[1]

The study ran for one week. None of the 689,003 people knew they were participants in a psychological experiment. None of them were asked. None of them consented in the way that any university research ethics board would require. At a traditional research institution, a study involving human subjects must pass through an Institutional Review Board before it begins. The board evaluates whether the study poses risks to participants. The board requires informed consent. The board ensures that subjects understand what is being done to them before they agree to participate. These safeguards exist because the history of human experimentation contains enough horrors to justify extreme caution. Facebook bypassed all of them. The company did not need consent because its Terms of Service contained a clause permitting the use of data for "internal operations, including troubleshooting, data analysis, testing, research, and service improvement." Somewhere in the thicket of legal language that no reasonable person reads in full, the company had secured permission to manipulate the emotional lives of its users. The experiment was legal. The experiment was also, by any ordinary moral standard, an act of profound invisibility. A company had reached into the inner lives of hundreds of thousands of people and rearranged them. The people never felt the hand.

Two years later, in June of 2014, the researchers published their findings in the Proceedings of the National Academy of Sciences.[1] The paper was titled "Experimental Evidence of Massive-Scale Emotional Contagion Through Social Networks." It was the publication that drew the outrage. Ethicists and psychologists objected. Editorial boards debated. The public recoiled at the idea that their emotions had been treated as variables in an equation they never saw. Facebook issued a careful statement acknowledging that "the research benefits of the paper may not have justified all of this anxiety." The language was telling. The company did not apologize for conducting the experiment. The company expressed regret that people found out about it.

This is the detail that matters most. The backlash occurred because the researchers chose to publish. They chose to pull back the curtain. The experiment itself had been invisible. If no one had written a paper about it, no one would have known. The 689,003 people would have gone on believing that their moods were their own. They would have attributed their slightly darker or slightly brighter week to the ordinary rhythms of life. The manipulation would have been perfect in its invisibility.

The question that follows is the one that should keep us awake at night. If this particular experiment only became public because researchers wanted academic credit for their work... how many experiments were conducted and never published? How many dials were turned that no one ever reported? Facebook had the capacity to manipulate the emotional states of its entire user base. The company had the legal cover to do so. The only constraint was the decision to tell someone about it afterward. That constraint proved to be entirely optional.

Facebook built something powerful. The company built a ring that made its actions invisible while keeping its users completely visible. Every click, every pause, every late-night scroll through a feed at two in the morning was observed and recorded and analyzed. The watchers remained unseen. The watched never knew they were being watched. This was a new geometry of power that the ancient philosophers never imagined.

On October 5, 2021, a woman named Frances Haugen sat before the United States Senate Committee on Commerce, Science, and Transportation. She was thirty-seven years old. She had studied electrical and computer engineering at Olin College and earned a master's degree in business from Harvard. She had worked at Google, at Pinterest, at Yelp. She was not a disgruntled employee with a personal grievance. She was a data scientist with a conscience. She had joined Facebook in 2019 specifically to work on the company's civic integrity team, the group responsible for addressing misinformation and the manipulation of elections on the platform. She believed the work mattered. She believed Facebook could be made better. Over the next two years, she watched the company make decisions that contradicted everything the civic integrity team was trying to accomplish. Before her testimony, she had copied tens of thousands of internal documents and delivered them to the Wall Street Journal, which published them as a series called "The Facebook Files."[2],[3]

Haugen was calm and precise in her testimony. She did not speak in the language of outrage or moral condemnation. She spoke in the language of corporate metrics and internal research memos and product decisions that she had witnessed from the inside. She presented data. She cited internal slide decks by their titles. She described meetings she had attended and decisions she had watched being made. The picture she painted was devastating not because it was speculative but because it was documented in the company's own words. Facebook's own researchers had produced the evidence. Facebook's own leadership had chosen to bury it. The documents Haugen leaked were not the observations of an outsider guessing at what happened behind closed doors. They were the company's own record of what it knew and when it knew it.

The first revelation concerned Instagram and teenage girls. Facebook's internal research had found that Instagram made body image issues worse for one in three teenage girls who already struggled with their self-perception.[3] The company's own slides stated the finding plainly. "We make body image issues worse for one in three teen girls." Another internal presentation noted that "teens blame Instagram for increases in the rate of anxiety and depression." The research was thorough. The research was also secret. Facebook did not publish it. Facebook did not share it with parents or regulators or mental health professionals. The company continued operating Instagram without changes while knowing what its own data showed. When members of Congress asked Mark Zuckerberg directly whether the platform harmed children, he offered carefully hedged answers that did not reflect the clarity of the internal findings.

The second revelation was a system called XCheck.[3] The name was short for "cross-check." On its surface, XCheck was described internally as a quality control measure. In practice, it was a secret program that exempted high-profile users from Facebook's standard content moderation rules. Celebrities, politicians, journalists, and other prominent accounts were flagged in the system. When they posted content that violated the platform's rules, their posts were not removed through the normal automated process. Instead, the posts were routed to a special review queue where they often lingered without action. In many cases, the posts remained online indefinitely. Facebook had created a two-tier system of justice on its platform. Ordinary users faced automated enforcement. Powerful users operated under a different set of rules. The public never knew. The system was invisible by design.

The third revelation concerned the algorithm itself. Facebook's internal researchers had studied the effects of the company's engagement-based ranking system. They found that the algorithm systematically promoted content that generated outrage and division.[3] Angry content produced more comments, more shares, more reactions. The algorithm read those signals as indicators of value. It promoted the content further. Facebook's researchers proposed specific changes to reduce the amplification of inflammatory material. Company leadership rejected those proposals because the changes would reduce engagement metrics. The company chose its numbers over its stated values. The choice was made in private. The billions of people whose information diets were shaped by the algorithm never saw the internal debate or the decision.

The fourth revelation traced back to a specific moment in 2018 when Mark Zuckerberg announced what he called the "Meaningful Social Interactions" pivot. The public-facing message was warm and reassuring. Facebook would prioritize content from friends and family over content from publishers and brands. The company would foster genuine human connection. Internally, the reality was different. The "Meaningful Social Interactions" metric rewarded content that generated comments and reactions. Facebook's own researchers quickly identified the problem. The content that generated the most comments and reactions was not heartwarming or genuinely social. It was inflammatory, divisive, and outrage-inducing. The metric was doing the opposite of what the public announcement promised. Researchers flagged this. Leadership was informed. The metric remained in place.[3]

Haugen's testimony before the Senate distilled the pattern into a single sentence. "The company's leadership knows how to make Facebook and Instagram safer but won't make the necessary changes because they have put their astronomical profits before people."[2] The statement was striking not for its anger but for its precision. Haugen was not speculating about corporate motives. She was describing documented decisions made in documented meetings by identified individuals. The evidence existed in the company's own files. The decisions had been made behind closed doors. They affected the information environment of roughly three billion people who used Facebook's products. Three billion people lived inside an architecture shaped by choices they could not see, made by people they could not hold accountable, for reasons they were never told.

Guillaume Chaslot understood the machinery from the inside. He was a French computer scientist with a doctorate in artificial intelligence who had joined YouTube's engineering team to work on the recommendation algorithm. The algorithm was YouTube's most important product. It decided what users watched next. When a video ended, the algorithm selected the next suggestion. When a user opened the app, the algorithm arranged the homepage. The recommendation engine drove roughly seventy percent of all viewing time on the platform. It was the invisible hand that guided billions of hours of human attention every day.[4] Chaslot had believed he was building something useful. He was helping people find videos they would enjoy. The math was elegant. The scale was exhilarating. But as he studied the system's behavior more closely, he began to notice patterns that troubled him. The algorithm was not simply helping people find what they wanted. The algorithm was steering people toward content they had never sought, content that was more extreme, more conspiratorial, more emotionally charged than what they had started with. The system optimized for engagement, and engagement correlated with outrage.

Chaslot left YouTube and dedicated himself to studying the system he had helped build. He created a project called AlgoTransparency, hosted at algotransparency.org, which tracked what YouTube's algorithm recommended to users.[4] His methodology was straightforward. He ran automated searches on various topics and followed the chain of recommendations to see where the algorithm led. The findings were consistent and disturbing. The algorithm systematically pushed users toward more extreme content. A user who searched for information about vaccines would find anti-vaccination content in the recommendations within a few clicks. A user watching mainstream political commentary would be steered toward increasingly radical material. This tendency cut across topics and across political orientations. The algorithm did not have a political agenda. It had an engagement agenda. Extreme content held attention longer than moderate content. The algorithm optimized for watch time. Extremism was good for watch time.

YouTube's own internal researchers identified the same pattern. They documented the radicalization pipeline. They proposed changes to the recommendation system that would reduce the amplification of extremist content. Management weighed the proposals against their potential impact on "watch time," which was YouTube's core performance metric. The changes were delayed or diluted or shelved.[4] The algorithm continued to operate as designed, steering hundreds of millions of users toward content that the company's own researchers had flagged as harmful. The users never saw the internal debates. The users only experienced the recommendations, which arrived with the seamless authority of a system that seemed to know what they wanted better than they knew themselves.

On March 15, 2019, a gunman walked into two mosques in Christchurch, New Zealand, and murdered fifty-one people during Friday prayers. He livestreamed the massacre on Facebook using a helmet-mounted camera.[9] The video was broadcast in real time to a relatively small audience. Then the algorithm took over. The video spread across the platform faster than the company's moderation systems could contain it. Users downloaded the footage and re-uploaded it. They clipped it into shorter segments. They altered it slightly to evade automated detection systems. In the first twenty-four hours, Facebook blocked 1.5 million attempts to upload the video. Many more copies circulated successfully. The footage migrated to YouTube, to Twitter, to Reddit, to smaller platforms with fewer moderation resources. It was shared, re-uploaded, embedded in memes, edited into propaganda.[9] The New Zealand Royal Commission of Inquiry later found that the attacker had specifically chosen to livestream on Facebook because he understood the platform's architecture.[9] He knew that the design would amplify his act. He knew that the algorithms would spread the footage faster than any human moderator could contain it. He had studied the platforms the way a general studies terrain before a battle. The livestream was not incidental to the attack. The livestream was part of the attack. The platform's own machinery had been weaponized by someone who understood its invisible logic better than its users did.

The Christchurch shooting exposed a truth that the platforms had worked to obscure. The algorithms that shaped what billions of people saw were not neutral pipes delivering information. They were active systems that made choices about amplification and suppression. Those choices were invisible to the public. The code was proprietary. The logic was opaque. No one outside the companies could examine the recommendation algorithms or audit the content moderation systems or verify the claims that the platforms made about their own safety measures. An algorithm that shaped the worldview of billions operated in complete darkness. The people inside the companies who understood how the systems worked were bound by employment agreements that prohibited disclosure. The people outside the companies who were affected by the systems every day had no way to see inside. This was the architecture of one-directional invisibility. The algorithm saw everything about its users. The users saw nothing about the algorithm.

The concept needs a name. Call it asymmetric visibility. It describes a power arrangement that Glaucon never considered when he told the story of the Ring of Gyges. Glaucon imagined a single person who became invisible among visible people. The scenario was symmetrical in one important sense. The invisible person and the visible people occupied the same world, subject to the same physical laws. The invisible person simply had one advantage that the others lacked. If the ring had been dropped, anyone could have picked it up. The advantage was portable, transferable, accidental. The modern technology platform operates under a different logic entirely. The invisibility is not a personal advantage that one individual holds over another. The invisibility is an institutional condition embedded in corporate structures, legal frameworks, and technical architectures that are designed to resist transparency. No individual can simply pick up the ring. The ring is the system itself.

Consider what the platform knows about you. It knows your location at this moment, tracked by the GPS in your phone. Your browsing history across hundreds of websites has been captured by tracking cookies and embedded pixels. Your purchase behavior is collected from credit card transactions and retail partnerships. It maps your social connections in granular detail through your contacts, your interactions, and the interactions of people who know you. Your communication patterns are logged as well, including who you talk to, when you talk to them, how long you talk, and what you say. If you have used the platform's photo features, it knows your face. If you have used its voice features, it knows your voice. Through inference models built on behavioral data, the platform can estimate your emotional state, your political orientation, your health status, and your financial situation.[8] You are, to the platform, a transparent object. Every dimension of your life is visible.

Now consider what you know about the platform. You know almost nothing. The algorithm that determines what you see is proprietary. The data collection practices are disclosed in Terms of Service documents that average tens of thousands of words and are written in language designed to inform lawyers, not users. The content moderation rules are applied by a combination of automated systems and human reviewers operating under guidelines that shift constantly and are never fully published. The advertising system that determines which messages are targeted at you and why operates on logic that is not available for your inspection. You exist inside a system that knows everything about you while telling you nothing about itself.[8]

Amazon's warehouse operations offer a concrete illustration of this asymmetry in its most extreme form.[5] Warehouse workers at Amazon fulfillment centers are tracked by the second. Their scanning rates are monitored. Their bathroom breaks are timed. Their movement patterns through the facility are recorded. Algorithms evaluate their productivity and flag deviations from expected performance. Workers who fall below algorithmically determined thresholds receive automated warnings. Workers who continue to fall below those thresholds are terminated, sometimes by automated systems that generate the termination notice without direct human involvement. The worker is completely visible to the system. Every motion is measured. Every pause is noted. Every deviation is recorded. Reports from warehouse workers describe an environment of constant surveillance where even a trip to the restroom is weighed against a productivity clock that never stops running.[5] The human body becomes a set of data points, measured against benchmarks that were set by an algorithm the worker will never see and can never question.

Meanwhile, the corporate decisions that govern the warehouse operate behind near-total secrecy. Amazon's internal deliberations about productivity targets, about acceptable injury rates, about the algorithms that evaluate worker performance are among the most closely guarded secrets in the technology industry.[5] Workers cannot see the algorithm that evaluates them. They cannot examine the logic that determines whether their scanning rate is fast enough. They cannot challenge the system that decides their employment with any understanding of how that system works. The same company that demands total visibility from its lowest-paid workers insists on total invisibility for itself. The asymmetry is not incidental. The asymmetry is the design.

This arrangement is fundamentally different from the scenario that Glaucon described. Glaucon imagined a symmetry of possibility. One person among equals happened to acquire an extraordinary advantage. The ring could theoretically have been found by anyone. The modern platform creates a permanent structural division between two classes of people. One class is visible at all times. The other class is invisible by structural design. The visible class cannot acquire the ring because the ring is not an object. The ring is an infrastructure that costs billions of dollars to build and requires thousands of engineers to maintain. The invisible class has no incentive to relinquish its advantage because its entire business model depends on maintaining the asymmetry. This is not the thought experiment of a philosopher exploring human nature. This is the operating model of the most powerful companies in human history.

The invisibility of algorithmic decision-making extends far beyond social media feeds and warehouse productivity scores. It reaches into the domains where the stakes are highest and the need for transparency is most urgent.

In 2018, Reuters reported that Amazon had built an artificial intelligence recruiting tool designed to automate the review of job applications.[7] The system was trained on ten years of hiring data. It learned to evaluate resumes by studying the patterns in the resumes of people Amazon had previously hired. The problem was that Amazon had previously hired mostly men, particularly in technical roles. The AI learned to penalize resumes that contained the word "women's," as in "women's chess club captain" or "women's studies." It downgraded graduates of all-women's colleges. The system had absorbed the biases embedded in the historical data and automated them at scale. Amazon scrapped the tool when internal testing revealed the problem.[7] The company deserves a measure of credit for that decision. The deeper question is how many similar systems at other companies were never tested for bias and never scrapped. The deeper question is how many applicants were rejected by invisible algorithms that carried the prejudices of their training data into decisions about human livelihoods.

In May of 2016, the investigative journalism organization ProPublica published an investigation into a criminal justice risk assessment algorithm called COMPAS.[6] The Correctional Offender Management Profiling for Alternative Sanctions system was used by courts across the United States to assess the likelihood that a defendant would commit future crimes. Judges consulted the algorithm's scores when making decisions about bail, sentencing, and parole. ProPublica's analysis found that the system was biased against Black defendants.[6] Black defendants were almost twice as likely as white defendants to be falsely flagged as future criminals. White defendants were almost twice as likely as Black defendants to be incorrectly labeled as low risk when they actually went on to commit future offenses. The algorithm was wrong in both directions, and both errors fell along racial lines.

Northpointe, the company that built COMPAS, disputed ProPublica's methodology and conclusions. Academic researchers debated the statistical definitions of fairness and whether any algorithm could satisfy all reasonable definitions simultaneously. The technical debate was important. It was also, in a sense, beside the point. The fundamental problem was that courts were using a proprietary algorithm to make decisions about human freedom. A judge in Wisconsin could consult a COMPAS score before deciding whether to sentence a defendant to prison or probation. The defendant had no right to examine the model that produced the score. The defendant's attorney could not cross-examine the algorithm. The defense could not challenge the reasoning because the reasoning was a trade secret owned by a private corporation. The most consequential decision in a person's life... whether to walk free or be locked in a cage... was influenced by a system that operated behind a wall of corporate confidentiality.[6] The ring was not on the finger of a shepherd. The ring was embedded in the code of a for-profit software company, and it was helping decide who went to prison.

The same asymmetry extends across domains. Credit scoring algorithms determine who receives loans and at what interest rates. A person denied a mortgage may never learn that the decision was made by an automated system trained on data that encodes decades of discriminatory lending practices. Insurance algorithms determine who pays more for coverage and who is denied entirely. A driver charged a higher premium may never discover that the algorithm factored in their zip code as a proxy for race. Content moderation algorithms determine what speech is permitted on the platforms where public discourse now takes place. A user whose post is removed may never see the rule that was applied or understand why the same rule was not applied to an identical post by someone else. Each of these systems makes consequential decisions about human lives. Each of these systems operates with minimal transparency. The people affected by the decisions rarely know that an algorithm made the decision. They almost never understand how the algorithm reached its conclusion. They have limited ability to challenge the outcome because the system that produced the outcome is hidden from them.

The European Union attempted to address this problem through the General Data Protection Regulation, which took effect in 2018.[10] The GDPR includes provisions that give individuals the right to receive meaningful information about the logic involved in automated decision-making. Article 22 gives individuals the right not to be subject to decisions based solely on automated processing when those decisions produce significant effects.[10] The principle is sound. The ambition is admirable. The enforcement has been limited. Technology companies have argued that full algorithmic transparency would expose trade secrets and harm their competitive position in global markets. They have deployed legal teams to narrow the interpretation of the regulations. They have spent millions lobbying against stronger enforcement mechanisms. They have complied with the letter of the law in ways that do not always serve the spirit of it. A user who exercises their right to an explanation may receive a response that is technically compliant but practically meaningless, a document that describes the categories of data considered without revealing how those categories were weighted or combined. The explanation explains nothing. The transparency is opaque. The right exists on paper. The power remains behind the curtain.

The fundamental problem persists. Decisions that shape billions of lives are made by systems that no one outside the company fully understands. In many cases, no one inside the company fully understands them either. Modern machine learning systems operate through patterns that emerge from training data in ways that their designers cannot always explain or predict. The engineers who build the systems can describe the architecture. They can identify the training data. They can measure the outputs. They cannot always articulate why the system produces a particular result in a particular case. The algorithm has become a kind of autonomous entity, invisible not because someone chose to hide it but because its internal logic is opaque even to its creators. The ring has become something that Plato could not have anticipated. It is a ring that no one wears because no one needs to. The invisibility is built into the structure itself.

The technology platform is the most powerful ring in the history of human civilization. Consider the properties. The ring sees everything. Every search, every click, every message, every purchase, every hesitation, every late-night confession typed into a search bar at three in the morning is collected and stored and analyzed. The data does not expire. The data does not forget. A search you conducted ten years ago still exists in a database somewhere, linked to your identity through a web of cookies and device identifiers and email addresses that trace a continuous line from the person you were then to the person you are now. The platform knows what you want before you articulate it. The platform knows what you fear before you admit it. The platform knows what you will buy before you decide to buy it. This is not metaphor. This is the operational reality of surveillance capitalism, the term coined by Shoshana Zuboff to describe the economic logic that drives the modern technology industry.[8] Zuboff describes a system in which human experience is claimed as free raw material for translation into behavioral data. The data is then processed into predictions of human behavior. The predictions are sold to business customers in what Zuboff calls "behavioral futures markets."[8]

The ring is seen by no one. The algorithms are proprietary. The data practices are buried in legal documents. The internal research is confidential. The decision-making processes are shielded by corporate secrecy. When the evidence of harm becomes public, it becomes public only through leaks and whistleblowers who risk their careers to speak. Frances Haugen lost her job. Guillaume Chaslot left his. The system does not volunteer transparency. Transparency must be extracted from it, at personal cost to the people who do the extracting.

The ring shapes behavior. The recommendation algorithm determines what information reaches you. Which voices you hear depends on the ranking system. The notification system governs when your attention is captured and for how long. And the advertising engine decides which desires are planted in your consciousness. Billions of people make daily decisions about what to believe, what to buy, what to fear, who to trust, based on information that was curated for them by invisible systems optimizing for engagement metrics. The platforms do not merely reflect the world. The platforms construct the world that their users inhabit.

The ring is not shaped by accountability. Facebook's own research on Instagram's harm to teenage girls, discussed earlier in this chapter, demonstrated the same willingness to prioritize engagement over safety. YouTube's internal findings about its radicalization pipeline, also documented above, followed the same logic of shelving reforms that threatened watch time. When the COMPAS algorithm was shown to be biased against Black defendants, the company that built it disputed the findings rather than opening the system for independent audit. When Amazon's hiring algorithm was caught discriminating against women, the company scrapped that particular tool but continued to use algorithmic systems for other hiring decisions without public disclosure of their design. The same dynamic recurs across industries and across decades. When the evidence of harm emerges, the response is denial, deflection, or delay. The invisible hand does not welcome the spotlight. Accountability requires visibility. The ring provides invisibility. The two are structurally incompatible.

The ring profits from attention while externalizing the costs of the attention economy. This is the economic logic that makes the ring self-sustaining. The platforms capture the revenue generated by user engagement. The advertising dollars flow to the companies. The mental health consequences fall on the users. The political polarization falls on the society. The erosion of shared reality falls on the democracy. The profits are private. The costs are public. The ledger is hidden. The platforms have constructed a business model that privatizes the benefits of mass attention while socializing the harms. This is not an accusation of malice. It is a description of incentive structures operating behind a curtain of invisibility. The people making these decisions may genuinely believe they are building tools that connect the world and democratize information. The invisibility of the consequences allows them to maintain that belief.

Glaucon's ancient question found its most powerful expression yet in the algorithm. The technology platforms have provided a modern answer to that question. The answer is not theft in the traditional sense. The answer is something more sophisticated and more totalizing. The answer is monetization. The platforms took the ring and used it to build an economy of invisible observation. They watched what people did when people thought no one was watching. They cataloged the private behaviors and the secret desires and the unguarded moments. They built profiles of such depth and precision that advertisers could target a person not just by their demographics but by their emotional state at a particular moment on a particular day. They sold access to that catalog to companies who wanted to reach people at their most vulnerable and most persuadable. A woman searching for information about pregnancy tests at midnight could find herself seeing advertisements for prenatal vitamins by morning, before she had told anyone in her life. A man researching symptoms of depression could find his feed populated with pharmaceutical advertisements within hours turning the visible population into a product.

The ancient ambition was straightforward... power, status and control. The technology platforms used a different kind of ring to pursue a different kind of throne. They did not overthrow governments, although their platforms destabilized some. They did not seduce queens, although their algorithms exploited desire at industrial scale. They built something that Gyges never imagined. They built a throne that did not require a visible ruler. The power operated through systems rather than through persons. The control was exercised through algorithms rather than through commands. The invisibility was not a cloak worn by an individual. The invisibility was the architecture itself.

The curtain remains in place. Behind it, the algorithms sort and rank and recommend and suppress, performing billions of calculations every second across server farms that consume more electricity than some nations. Behind it, the data flows into models that predict and manipulate, growing more sophisticated with every interaction they observe. Behind it, the decisions are made about what three billion people will see when they open their phones in the morning. The curtain is not a failure of the system. The curtain is the system. It was designed to be opaque. It profits from its opacity. Every attempt to pull the curtain back has been met with resistance, with legal challenges, with appeals to trade secrets and competitive advantage. The companies that built the most powerful surveillance apparatus in human history have wrapped it in the language of connection, community, and free expression. The language is visible. The machinery is not.

In Plato's myth, the ring was discovered by accident, its power neither earned nor designed. The modern version of the ring was not found by accident. It was engineered deliberately by some of the most talented minds in the history of technology. They built it with purpose. They built it with precision. They built it to do exactly what it does. The engineers were not villains in the storybook sense. Many of them believed they were building something that would bring the world closer together. Many of them still believe that. The systems they constructed were not designed to manipulate or exploit. The systems were designed to optimize. The distinction matters because it reveals something important about the nature of the modern ring. The harm is not always intentional. The harm is structural. The optimization function does not account for the cost of what is being optimized. The engagement metric does not measure the anxiety it generates. The watch time counter does not track the radicalization it enables. The system pursues its objective with perfect efficiency while remaining perfectly blind to the damage it creates along the way. The question is not whether the people who built these systems will use the power that invisibility provides. They already have. The question is whether the rest of us will ever be allowed to see what they are doing with it... or whether the curtain will hold.

References

[1] Kramer, Adam D.I., Jamie E. Guillory, and Jeffrey T. Hancock. "Experimental Evidence of Massive-Scale Emotional Contagion Through Social Networks." *Proceedings of the National Academy of Sciences* 111, no. 24 (2014): 8788-8790.

[2] Haugen, Frances. Testimony before the U.S. Senate Committee on Commerce, Science, and Transportation. October 5, 2021.

[3] Horwitz, Jeff, et al. "The Facebook Files." *Wall Street Journal,* September-October 2021 (series).

[4] Chaslot, Guillaume. Research and documentation at algotransparency.org.

[5] Rosenblat, Alex. *Uberland: How Algorithms Are Rewriting the Rules of Work*. University of California Press, 2018.

[6] Angwin, Julia, et al. "Machine Bias." *ProPublica*, May 23, 2016.

[7] Dastin, Jeffrey. "Amazon Scraps Secret AI Recruiting Tool That Showed Bias Against Women." *Reuters*, October 10, 2018.

[8] Zuboff, Shoshana. *The Age of Surveillance Capitalism: The Fight for a Human Future at the New Frontier of Power*. PublicAffairs, 2019.

[9] Royal Commission of Inquiry into the Attack on Christchurch Mosques. Report, November 2020.

[10] European Union. General Data Protection Regulation (GDPR), Articles 13-15, 22. 2016/2018.

Chapter Ten - What Governments Do in the Dark

In the spring of 1953, a thirty-seven-year-old biochemist named Frank Olson drove from his home in Frederick, Maryland, to a government research facility at Camp Detrick. He had done this commute hundreds of times before. Olson worked for the United States Army's Special Operations Division, where he studied biological weapons and the methods by which they might be deployed. He held a top-secret clearance. He was trusted with some of the most sensitive research in the American defense establishment.

On November 19 of that year, Olson and several colleagues gathered at a CIA retreat in Deep Creek Lodge, a rustic cabin in western Maryland. Sidney Gottlieb, the head of the CIA's Technical Services Staff, was there to discuss new research initiatives. At some point during the evening, Gottlieb spiked a bottle of Cointreau with lysergic acid diethylamide. He poured glasses for the unsuspecting scientists. Twenty minutes later, he told them what he had done.[1]

Frank Olson's world came apart over the following ten days. He became agitated, paranoid, and deeply depressed. His colleagues noticed the change immediately. His supervisor contacted the CIA, which arranged for Olson to see a physician in New York City. On the night of November 28, 1953, Olson plunged from the thirteenth-floor window of the Statler Hotel on Seventh Avenue. He was found on the sidewalk below, wearing only his underwear. The New York City medical examiner ruled the death a suicide.[2]

His family accepted this explanation for twenty-two years. They believed Frank Olson had suffered a breakdown and taken his own life. They mourned him and moved on with their lives as best they could. Then, in 1975, a congressional investigation revealed that Olson had been dosed with LSD without his knowledge as part of a secret CIA program called MKULTRA. The family learned that everything they had been told about Frank Olson's death was either incomplete or false.[3]

In 1994, after exhuming his body, a forensic pathologist found a cranial injury consistent with a blow to the head before the fall. The New York district attorney opened a homicide investigation but never brought charges. To this day, the Olson family does not know whether Frank Olson jumped from that window, fell, or was pushed. They know only that the Central Intelligence Agency drugged him without his consent, watched him deteriorate, and then offered an account of his death that concealed the agency's role in causing it.[2]

Frank Olson's story is not an anomaly. It is a data point in a pattern so consistent that it functions as a kind of natural law. When governments operate in secret, shielded by classification stamps, national security claims, and sovereign immunity, they behave the way Glaucon predicted any person would behave with a ring of invisibility. They do things they would never do if the public could see them. The Ring of Gyges does not merely tempt individuals. It tempts entire systems of power.

This chapter documents four cases spanning seven decades of American governance. Each case is different in its specifics. Each involved different agencies, different administrations, and different political contexts. Yet the underlying structure is identical in every instance. A government agency operated in secret. It did things far beyond what the public would have tolerated. The secrecy was eventually broken not by internal oversight but by accident, theft, or the conscience of a single person. When the truth emerged, the institution shielded its leaders and punished those at the bottom of the hierarchy. Reforms followed, and then the cycle began again.

This is not a partisan argument. These programs were authorized by presidents of both parties, funded by congresses of both parties, and concealed from the public by bureaucracies that served both parties. The ring does not care about ideology. It cares only about invisibility.

The program that killed Frank Olson was called MKULTRA, and its ambitions were staggering. In the early years of the Cold War, American intelligence officials became convinced that the Soviet Union and China had developed techniques for brainwashing prisoners of war. American soldiers captured during the Korean War appeared on Communist propaganda broadcasts confessing to crimes they had not committed, denouncing their own country in language that sounded rehearsed and mechanical. The CIA concluded that the Soviets possessed some form of mind control technology. The agency wanted to find it, understand it, and build its own version.[1]

On April 13, 1953, CIA Director Allen Dulles authorized the creation of MKULTRA under the supervision of Sidney Gottlieb, a chemist with a club foot and a passion for folk dancing who would become one of the most consequential figures in the history of American intelligence. Gottlieb's mandate was broad. He was to explore any method that might allow the agency to control the human mind. The word "any" was interpreted literally.[3]

Over the next twenty years, MKULTRA grew to encompass at least 149 separate research projects across more than eighty institutions, including universities, hospitals, pharmaceutical companies, and prisons. The methods ranged from the merely unethical to the genuinely sadistic. CIA researchers administered LSD to unwitting subjects at bars, restaurants, and brothels in an operation called Midnight Climax, where sex workers hired by the agency lured men to safe houses and the researchers watched through one-way mirrors as the drugs took effect. They tested mescaline, barbiturates, amphetamines, and heroin. They experimented with electroshock therapy, sensory deprivation, and prolonged isolation. They studied hypnosis, subliminal messaging, and the psychological effects of sustained verbal abuse.[1]

Some of the most devastating work took place at the Allan Memorial Institute in Montreal, Canada. There, a Scottish-born psychiatrist named Dr. Donald Ewen Cameron conducted experiments that he called "psychic driving." Cameron believed he could erase a patient's existing personality and rebuild it from scratch. His methods involved placing patients in drug-induced comas for weeks at a time while playing looped audio messages through speakers beneath their pillows. He administered massive doses of electroconvulsive therapy, sometimes dozens of sessions over a few days, at voltages far exceeding anything used in legitimate medical practice. He combined these treatments with sensory deprivation, keeping patients in darkened rooms and depriving them of contact with other human beings.[3]

Cameron's patients had come to the Allan Memorial Institute seeking treatment for depression, anxiety, and other common psychiatric conditions. They did not know they were experimental subjects. They did not know the CIA was funding the research. Many of them emerged from Cameron's care with shattered memories and destroyed personalities. Some could no longer recognize their own families. Others spent the rest of their lives unable to function independently. The Canadian government eventually paid compensation to some of the victims, but the damage was irreversible. Cameron himself was never prosecuted. He died of a heart attack while mountain climbing in 1967, still regarded as one of the most distinguished psychiatrists in North America.[3]

The full scope of MKULTRA might never have become public. In 1973, as congressional investigations into intelligence abuses began to gather momentum, CIA Director Richard Helms ordered the destruction of all MKULTRA files. His subordinates complied with thorough efficiency. Thousands of documents detailing two decades of secret experimentation on American citizens and foreign nationals were fed into shredders. The program's history was meant to disappear entirely.[1]

It nearly did. The only reason any records survived was a filing error. Approximately 20,000 financial documents related to MKULTRA had been stored in a separate archive because they pertained to budgetary rather than operational matters. When Helms ordered the destruction of the files, nobody thought to check the financial records office. In 1977, a

Freedom of Information Act request by investigative journalist John Marks uncovered these surviving documents, and the outline of the program finally became visible.[1]

Consider what this means. An intelligence agency of the United States government conducted mind-control experiments on unwitting human subjects for twenty years. The experiments caused permanent psychological damage to an unknown number of people and contributed to at least one death. The agency then attempted to erase all evidence that the program had ever existed. It failed only because someone misfiled a box of receipts. The ring of Gyges operated at the highest level of American power, and it very nearly succeeded in making everything it touched disappear.

While the CIA was experimenting on the minds of its citizens, the Federal Bureau of Investigation was working to control their politics. On August 25, 1956, FBI Director J. Edgar Hoover authorized a secret program called COINTELPRO, an abbreviation for Counterintelligence Program. The stated purpose was to "expose, disrupt, misdirect, discredit, or otherwise neutralize" organizations that Hoover considered threats to national security. In practice, COINTELPRO targeted American citizens exercising their constitutional rights to free speech, free assembly, and political organizing.[4]

The program's initial focus was the Communist Party USA, which by the mid-1950s was already a marginal organization with dwindling membership and negligible political influence. But COINTELPRO quickly expanded far beyond its original mandate. By the early 1960s, the FBI was conducting covert operations against civil rights organizations, antiwar groups, feminist movements, and Native American activists. The targets were not foreign agents or violent revolutionaries. They were American citizens engaged in lawful political activity. The FBI treated them as enemies of the state because they held views that J. Edgar Hoover found objectionable.[4]

The most thoroughly documented target was Dr. Martin Luther King Jr. Beginning in 1963, the FBI placed wiretaps on King's home and office phones. Agents bugged his hotel rooms across the country. They recorded his private conversations, his phone calls, and his intimate moments with women other than his wife. Hoover personally authorized these operations because he believed King was a dangerous radical influenced by Communist advisors. The surveillance was relentless and comprehensive. It continued until King's assassination in April 1968.[4]

In November 1964, the FBI sent King an anonymous package. Inside was a letter and an audio recording. The letter, which was not declassified until 2014, urged King to action. "King, there is only one thing left for you to do," the letter read. "You know what it is. You have just 34 days in which to do it." The letter threatened to release damaging personal information unless King removed himself from public life. The audio recording contained material the FBI had gathered through its illegal surveillance.[4]

The nation's chief law enforcement agency sent one of its own citizens an anonymous letter encouraging him to exit public life. The letter was composed by FBI agents, approved by FBI leadership, and mailed from FBI offices. It was kept secret for fifty years. King was far from the only target. The FBI infiltrated dozens of Black civil rights organizations with informants and provocateurs. Agents wrote fake letters designed to create suspicion and infighting within these groups. They planted false stories in newspapers. They pressured landlords to evict activists and employers to fire them. The program was systematic and bureaucratic. Field offices submitted detailed proposals for "disruption operations" to FBI headquarters, which reviewed and approved them through a formal chain of command.[4]

The consequences sometimes extended beyond harassment into violence. On December 4, 1969, a team of fourteen Chicago police officers raided the apartment of Fred Hampton, the twenty-one-year-old chairman of the Illinois chapter of the Black Panther Party. The officers fired approximately ninety rounds into the apartment. Hampton was shot twice in the head while lying in his bed. His bodyguard, Mark Clark, was also killed. The surviving occupants were arrested and charged with attempted murder of the police officers.[4]

The raid had been planned with intelligence provided by the FBI. An informant named William O'Neal had given the Bureau a floor plan of Hampton's apartment, including the location of Hampton's bed. A toxicology report later revealed that Hampton had been drugged with secobarbital, a powerful sedative, raising questions about whether he was even conscious when the shooting began. A federal investigation eventually concluded that the officers had fired between eighty-two and ninety-nine rounds, while only one shot could be attributed to the apartment's occupants.[4]

Fred Hampton was an American citizen. He was not charged with any crime at the time of his death. He was a political organizer. The FBI facilitated his killing because his politics were deemed threatening. COINTELPRO operated for fifteen years. During that time, it was never authorized by any court. It was never disclosed to Congress. It was never subjected to any form of public oversight. The program was ended not because the FBI developed a conscience or because internal review processes identified abuses. It was ended because of a burglary.

On the night of March 8, 1971, while most of America was watching the heavyweight boxing match between Muhammad Ali and Joe Frazier, a small group of antiwar activists calling themselves the Citizens' Commission to Investigate the FBI, broke into a two-man FBI office in Media, Pennsylvania. They carried away every file in the building. Over the following weeks, they mailed the stolen documents to journalists and members of Congress.[4]

The documents revealed the existence of COINTELPRO. They exposed the FBI's systematic campaign of surveillance and disruption against lawful political organizations. The disclosures led to the Church Committee investigations of 1975 and 1976, which documented the full scope of intelligence abuses across multiple agencies. The citizens who broke into that office were never caught. Their identities remained unknown for more than forty years, until several of them came forward publicly in 2014.[4]

Thirty-two years after the end of COINTELPRO, the pattern repeated itself in a different theater of American power. This time, the invisibility was geographic. The abuses occurred on the other side of the world. The victims were not American citizens but foreign detainees held in a military prison. The exposure came not through a congressional investigation or a journalist's inquiry but through something the perpetrators never imagined would become public... their own photographs.

Abu Ghraib was a sprawling prison complex twenty miles west of Baghdad. Under Saddam Hussein, it had been a site of mass execution and torture. After the American invasion of Iraq in 2003, the U.S. military took control of the facility and used it to house thousands of Iraqi detainees. Conditions were chaotic. The prison was overcrowded, understaffed, and under frequent mortar attack. Military police officers responsible for guarding the detainees had received minimal training in detention operations. Intelligence officers pressured them to "soften up" prisoners for interrogation.[5]

Between October and December of 2003, American soldiers at Abu Ghraib subjected Iraqi detainees to systematic abuse. Prisoners were stripped naked and forced into degrading poses. They were stacked in human pyramids. They were dragged across floors on leashes. They were threatened with attack dogs. They were subjected to simulated electrocution. They were sexually humiliated in ways that were calculated to exploit cultural and religious sensitivities. The abuse was not spontaneous or accidental. It was persistent, organized, and carried out with an air of casual amusement.[5]

The soldiers who committed these acts did something that, in retrospect, seems almost incomprehensible. They photographed themselves doing it. They posed for pictures next to hooded and naked prisoners. They smiled and gave thumbs-up signs. Specialist Lynndie England was photographed holding a leash attached to a naked man lying on the floor. Specialist Charles Graner was photographed standing behind a pyramid of naked detainees with his arms crossed and a grin on his face. The photographs were passed around on compact discs and shared among soldiers as though they were vacation snapshots.[5]

The photographs reveal something essential about the psychology of the ring. These soldiers did not behave like people who feared being caught. They behaved like people who genuinely believed that what they were doing would never be seen by anyone outside their closed world. They documented their own crimes because the documentation was, in their minds, private. Abu Ghraib existed in a space beyond accountability. The prison was far from the American public, hidden behind classification and military bureaucracy, located in a war zone where normal rules seemed suspended. The soldiers operated as if they were invisible. They acted accordingly.

In January 2004, Specialist Joseph Darby gave a copy of the photographs to the Army's Criminal Investigation Command. The military launched an internal investigation. Major General Antonio Taguba, assigned to investigate the 800th Military Police Brigade, produced a report documenting what he called "numerous incidents of sadistic, blatant, and wanton criminal abuses" that were "systemic and illegal."[5]

Three months later, CBS News broadcast some of the photographs on 60 Minutes. The New Yorker published an extensive report by Seymour Hersh. The world reacted with horror. The images became some of the most recognizable photographs of the twenty-first century, symbols of American hypocrisy and military overreach that damaged the country's standing for years to come. The accountability that followed revealed the second half of the pattern. Eleven soldiers were convicted of charges ranging from dereliction of duty to assault. Lynndie England received a three-year sentence. Charles Graner received ten years, the longest sentence of any of those convicted. These were the people in the photographs, the low-ranking enlisted soldiers who had committed the physical acts of abuse.[5]

No senior officer or civilian official was prosecuted. Brigadier General Janis Karpinski, who commanded the 800th Military Police Brigade, was demoted to colonel. She was the highest-ranking officer to face any consequence. The intelligence officers who had encouraged the "softening up" of detainees were not charged. The senior military and civilian leaders who had created the conditions and the culture in which such abuse could flourish were never held accountable. The institution protected those at the top and sacrificed those at the bottom.[5]

Philip Zimbardo, the psychologist who had conducted the Stanford Prison Experiment in 1971, served as an expert witness at the courts-martial. His testimony drew a direct line between Abu Ghraib and his own research. Zimbardo argued that the situation at the prison contained every element his experiment had identified as capable of transforming ordinary people into perpetrators of cruelty. Anonymity, deindividuation, permission from authority, dehumanization of the outgroup, and a complete absence of external oversight. The ring of invisibility had been placed around an entire facility, and the results were consistent with what his research predicted.[11]

The soldiers at Abu Ghraib were not sociopaths. Their psychological evaluations showed them to be unremarkable. They were ordinary Americans placed in an environment where they believed no one was watching, no one would know, and no one would care. A prison on the outskirts of Baghdad and a belief that the photographs would never leave the building.

The final case in this chapter concerns the most extreme power any government can exercise, the power to kill. On September 30, 2011, a Predator drone flying over the Jawf province of Yemen fired a Hellfire missile at a convoy of vehicles. The strike killed Anwar al-Awlaki, a forty-year-old dual citizen of the United States and Yemen who had become a prominent figure in al-Qaeda in the Arabian Peninsula. Al-Awlaki was an Islamic cleric who had used the internet to recruit and inspire terrorist attacks against the United States. He was connected to the Fort Hood shooting, the attempted bombing of a Detroit-bound airliner, and other plots. By any reasonable assessment, he was a dangerous man.[8]

He was also an American citizen. He had been born in New Mexico. He held a valid United States passport. Under the Fifth Amendment to the Constitution, he could not be deprived of life without due process of law. He had never been charged with a crime, never indicted by a grand jury, never tried before a court.

The legal authority for the killing came from a secret memorandum written by the Office of Legal Counsel at the Department of Justice. The memo, which remained classified for years after the strike, concluded that the president possessed the constitutional authority to order the targeted killing of an American citizen abroad if that citizen posed an imminent threat, if capture was not feasible, and if the operation was conducted in a manner consistent with the laws of war. The definition of "imminent" in the memo was stretched beyond its ordinary meaning. A threat did not need to be immediate to qualify as imminent. It needed only to be ongoing.[9]

Two weeks after the strike that killed Anwar al-Awlaki, a second drone strike in Yemen killed his sixteen-year-old son, Abdulrahman al-Awlaki.

Abdulrahman was an American citizen born in Denver, Colorado. He was not suspected of any involvement in terrorism. He was a teenager looking for his father. The strike that killed him targeted a group of people eating dinner at an outdoor restaurant. The Obama administration initially suggested that Abdulrahman was a twenty-one-year-old militant, a claim that was quickly disproven by his birth certificate and family records. Senior administration officials later offered varying and contradictory explanations. Robert Gibbs, a former Obama campaign spokesperson, suggested in a media appearance that the boy should have "had a more responsible father."[9]

No official has ever been held accountable for the killing of Abdulrahman al-Awlaki. No investigation has ever produced a public finding about why he was killed or who authorized the strike that took his life.

The drone program that produced these killings operated within a classified framework known informally as the "disposition matrix," a database maintained by the National Counterterrorism Center that tracked individuals targeted for capture or killing. The process by which names were added to this list was secret. The legal memos authorizing the program were secret. The criteria for targeting were secret. The strikes themselves occurred in countries where the United States was not formally at war, against individuals who were sometimes identified not by name but by patterns of behavior interpreted through surveillance data.[9]

The Bureau of Investigative Journalism, an independent nonprofit based in London, documented the cumulative toll. Between 2004 and 2018, American drone strikes killed thousands of people across Pakistan,

Yemen, Somalia, and Libya. Estimates of civilian casualties ranged from several hundred to several thousand, depending on the methodology used and the definition of "civilian" employed. The true numbers remained contested because the strikes occurred in remote areas where independent verification was difficult and the government's own casualty counts were classified.[8]

The drone program represents the ring of Gyges at its most extreme. The state exercised the ultimate power over human life. It did so in secret. It justified its actions through classified legal reasoning that the public was not permitted to read. It operated through a secret bureaucratic process that the public was not permitted to know existed. When the killing of American citizens raised constitutional questions, the courts declined to review the matter, ruling that the decisions fell within the political branches' authority over national security. Congress received limited briefings and imposed no meaningful constraints.[10]

The power to kill, exercised in secret, with secret legal justification, minimal oversight, and no public accountability.

Four cases. Seven decades. Different agencies, different administrations, different political parties, different historical contexts. The details vary enormously, from LSD experiments in the 1950s to drone strikes in the 2010s. Yet the structural pattern is so consistent that it begins to look less like a series of isolated scandals and more like a feature of how power operates when it believes itself unseen.

The first element of the pattern is secrecy itself. In every case, the state operated behind a wall of classification, national security privilege, or institutional opacity. MKULTRA was classified. COINTELPRO was classified. The abuse at Abu Ghraib took place behind the walls of a military prison in a war zone. The drone program's legal framework was classified. Secrecy was not incidental to these programs. It was their enabling condition. Without secrecy, none of them could have existed in the form they took.

The second element is escalation. In every case, the actions taken in secret were more extreme than anything the American public would have authorized through democratic processes. No Congress would have voted to fund mind-control experiments on unwitting citizens. No public would have endorsed sending anonymous letters to civil rights leaders. Secrecy did not merely conceal these programs. It liberated them from the constraints that democratic accountability would have imposed. Once the ring was on, the bearer did things that would have been unthinkable in the light.

The third element is the failure of internal oversight. In every case, the abuses were discovered not through the mechanisms designed to prevent them but through external disruption. MKULTRA was exposed because someone misfiled a box of financial records. COINTELPRO was exposed because activists burglarized an FBI office. Abu Ghraib was exposed because soldiers passed around photographs that eventually reached someone with a conscience. The drone program's legal reasoning was exposed through leaked memos and FOIA litigation. The watchdogs did not catch the wolves. The wolves were caught by accident, by theft, or by betrayal.

The fourth element is asymmetric accountability. In every case where individuals were punished, the punishment fell on those at the bottom of the hierarchy. Low-ranking soldiers were convicted at Abu Ghraib while generals and civilian officials remained untouched. CIA operatives who conducted MKULTRA experiments never faced prosecution, while the agency's director who ordered the destruction of evidence served out his career with his reputation intact. The institution reflexively shielded its leadership. Power protected itself. The ring, even once removed, continued to provide a measure of invisibility to those at the top.

The fifth element is the cycle of reform and repetition. After each exposure, reforms were enacted. The Church Committee led to the Foreign Intelligence Surveillance Act of 1978. Abu Ghraib led to new guidelines for detainee treatment. The Snowden disclosures led to the USA FREEDOM Act. These reforms were real and meaningful. They also proved impermanent. The lesson is not that reforms are useless. The lesson is that secrecy is corrosive, and any oversight mechanism that operates within the culture of secrecy will eventually be absorbed by it.

Taken together, these four elements describe something more fundamental than a series of policy failures. When accountability disappears, when the ring is placed on the hand of a bureaucracy, the institution behaves exactly as Glaucon predicted. It pursues its objectives without constraint. It defines its own limits. It treats the absence of external oversight as permission to do whatever it believes is necessary.

Plato himself anticipated this problem. In the Republic, he worried obsessively about who would guard the guardians. His solution was to strip his philosopher-kings of private property and family ties, to make them so transparent that corruption would be impossible. The solution was impractical then and remains impractical now. But the diagnosis was precise. Power that cannot be seen cannot be held accountable. Invisibility does not merely change what people do. It changes what institutions become.

The cases documented in this chapter span administrations led by both major American political parties. MKULTRA was authorized under Eisenhower and continued through Kennedy, Johnson, and into the Nixon era. COINTELPRO operated under four presidents across both parties. The Abu Ghraib abuses occurred during a Republican administration. The drone program expanded dramatically under a Democratic one. The ring is nonpartisan. It offers the same temptation to every hand that wears it.

And the temptation is always the same. It is the promise that secrecy will make difficult things easier. That classification will remove the messy complications of public debate. That national security, invoked as a talisman, will transform actions that would otherwise be illegal or immoral into actions that are necessary and justified. The promise is seductive because it is partly true. Secrecy does make certain things easier. It removes obstacles. It accelerates decision-making. It eliminates dissent. These are genuine operational advantages. They are also the precise conditions under which abuse becomes inevitable.

The question is not whether governments need some degree of secrecy. Of course they do. Intelligence operations, military planning, and diplomatic negotiations all require confidentiality. The question is what happens when secrecy becomes the default rather than the exception, when classification is used not to protect legitimate security interests but to shield the government from the embarrassment of its own actions.

The evidence suggests that institutions wearing the ring are, if anything, more dangerous than individuals wearing it. An individual's capacity for harm is limited by their physical power and their lifespan. An institution can cause harm at industrial scale, sustaining it across decades and administrations, distributing responsibility so widely that no single person can be held accountable. The CIA did not stop conducting unethical experiments when Sidney Gottlieb retired. The FBI did not stop surveilling political dissidents when J. Edgar Hoover died. The programs continued because they were not the products of individual malice. They were the products of institutional incentives operating in the absence of light.

References

[1] Marks, John. *The Search for the "Manchurian Candidate": The CIA and Mind Control*. Times Books, 1979. Marks's account, drawn from the 20,000 surviving MKULTRA financial documents recovered through his FOIA request, remains the definitive reconstruction of the program's scope, including the Gottlieb-directed LSD experiments and the Helms-ordered destruction of files.

[2] Albarelli, H.P., Jr. *A Terrible Mistake: The Murder of Frank Olson and the CIA's Secret Cold War Experiments*. Trine Day, 2009. See also the 1994 forensic exhumation findings by James Starrs, Professor of Law and Forensic Science at George Washington University, which identified cranial injuries inconsistent with a simple fall.

[3] U.S. Senate Select Committee on Intelligence. "Project MKULTRA, the CIA's Program of Research in Behavioral Modification." Joint Hearing before the Select Committee on Intelligence and the Subcommittee on Health and Scientific Research, August 3, 1977. See also Collins, Anne. *In*

the Sleep Room: The Story of the CIA Brainwashing Experiments in Canada. Lester & Orpen Dennys, 1988, for the Allan Memorial Institute experiments under Dr. Donald Ewen Cameron.

[4] U.S. Senate Select Committee to Study Governmental Operations with Respect to Intelligence Activities (Church Committee). "Intelligence Activities and the Rights of Americans," Book II, Final Report, April 26, 1976. See also Medsger, Betty. *The Burglary: The Discovery of J. Edgar Hoover's Secret FBI*. Alfred A. Knopf, 2014, for the account of the Media, Pennsylvania, break-in and the anonymous letter sent to Martin Luther King Jr.

[5] Taguba, Antonio M. "Article 15-6 Investigation of the 800th Military Police Brigade" (The Taguba Report), 2004. See also Hersh, Seymour M. "Torture at Abu Ghraib." *The New Yorker*, May 10, 2004; and Gourevitch, Philip, and Errol Morris. *Standard Operating Procedure*. Penguin Press, 2008.

[7] Foreign Intelligence Surveillance Court (FISC), annual reports to Congress, 1979-2019. The 99.97 percent approval rate is calculated from published statistics on applications submitted and denied.

[8] The Bureau of Investigative Journalism. "Drone Warfare" database, ongoing project tracking U.S. drone strikes in Pakistan, Yemen, Somalia, and Libya, available at thebureauinvestigates.com. See also Scahill, Jeremy. *The Assassination Complex: Inside the Government's Secret Drone Warfare Program*. Simon & Schuster, 2016.

[9] U.S. Department of Justice, Office of Legal Counsel. Memorandum regarding the targeted killing of Anwar al-Awlaki, released in redacted form in June 2014 following FOIA litigation by the ACLU. See also Mazzetti, Mark. *The Way of the Knife: The CIA, a Secret Army, and a War at the Ends of the Earth*. Penguin Press, 2013, for the killing of Abdulrahman al-Awlaki and the "disposition matrix."

[10] For the judiciary's refusal to review targeted killing decisions, see *Al-Aulaqi v. Obama*, 727 F. Supp. 2d 1 (D.D.C. 2010). For congressional oversight limitations, see the Senate Intelligence Committee's periodic reviews of the drone program's legal authorities under the 2001 Authorization for Use of Military Force.

[11] Zimbardo, Philip. *The Lucifer Effect: Understanding How Good People Turn Evil*. Random House, 2007. Zimbardo's analysis of Abu Ghraib as an expert witness drew directly on his findings from the 1971 Stanford Prison Experiment, identifying situational factors including anonymity, deindividuation, and the absence of external oversight.

Chapter Eleven - The All-Seeing Eye

In the oldest Greek tellings, the giant had one hundred eyes. His name was Argus Panoptes, and Hera assigned him a single task... to watch the maiden Io without ceasing. Argus was the perfect sentinel. Some of his eyes could close while others remained open, so that vigilance was continuous. He never slept in the way that mortals understood sleep. He was surveillance made flesh, a creature built for one purpose alone.

The goddess had good reason to be suspicious. Zeus had seduced Io and then transformed her into a white cow to conceal the affair. Hera, unconvinced by the disguise, claimed the cow as a gift and posted Argus as its guard. The giant took his station and watched with those restless eyes as Io grazed in a field near Mycenae. Day passed into night and back again. The eyes took shifts among themselves, blinking in rotation like signal lights along a distant tower. Nothing escaped him.[1]

Zeus could not tolerate this arrangement. He sent Hermes, the cleverest of the gods, to free Io from her watcher. Hermes did not come armed for battle. He did not come armored for force. He came disguised as a shepherd, carrying a set of reed pipes and a head full of stories. He sat beside the giant and began to play. The music was slow and wandering, the kind of melody that makes the mind drift. When Argus seemed interested, Hermes began to speak. He told a long story about the nymph Syrinx and the god Pan, a tale about pursuit and transformation and the making of the very pipes he carried. The story went on without urgency. It circled and doubled back on itself like a stream through a flat meadow.

One by one, the hundred eyes began to close. The music and the narrative worked together, each reinforcing the other, until vigilance gave way to drowsiness. When the last eye shut, Hermes drew his blade and killed the giant where he lay.[1]

The myth carries a lesson that the Greeks understood better than most civilizations that followed. Even perfect surveillance can be defeated. The weapon that destroyed Argus was not stronger than Argus. It was not faster or more powerful. It was simply more patient. Hermes defeated the hundred-eyed giant through narrative, through the control of attention. He decided what the watcher would focus on and kept him focused there until the watching stopped.

Centuries after the myth was first told, the governments of the world's most powerful democracies built their own version of Argus. They constructed agencies with tens of thousands of employees and budgets measured in billions. They wired listening stations across every continent and tapped the cables that carry the world's communications beneath its oceans. They designed systems that never sleep, that monitor billions of conversations each day, that watch constantly across time zones in a rotation that would have impressed even Hera. These agencies were built to see everything, and they were designed to be seen by no one. The watchers wore the ring.

The most powerful intelligence alliance in human history operated for decades before most people learned it existed. The alliance has no formal headquarters and no public spokesperson. Its name sounds clinical, Five Eyes. Its origins trace to a single classified agreement signed in the aftermath of the Second World War.

In 1946, representatives of the United States and the United Kingdom signed what became known as the UKUSA Agreement.[2] The pact formalized a relationship that had developed during the war, when American and British code-breakers worked side by side to crack Axis communications. The two nations had shared signals intelligence during the conflict out of necessity. After the war ended, they decided to keep sharing it out of ambition.

The original agreement bound two nations. Within a decade, it expanded to include three more... Canada, Australia, and New Zealand.[2] Each member nation maintained its own signals intelligence agency. The United States had the National Security Agency, headquartered at Fort Meade in Maryland behind fences and guard towers. The United Kingdom had the Government Communications Headquarters, operating from a massive circular building in Cheltenham that intelligence officers called "the Doughnut." Canada operated the Communications Security Establishment from a campus in Ottawa. Australia ran what is now called the Australian Signals Directorate. New Zealand contributed the Government Communications Security Bureau. Five nations, five agencies, five sets of eyes pointed outward at the rest of the world. Together they covered every hemisphere and nearly every time zone, forming a surveillance chain that followed the sun around the planet.

The arrangement was elegant in its legal architecture. Each nation's domestic law placed restrictions on spying against its own citizens. The American Constitution's Fourth Amendment protected Americans from unreasonable searches. British law contained similar protections for subjects of the Crown. These were serious legal barriers with real judicial consequences. The Five Eyes alliance offered a workaround that was as simple as it was effective. Each member agreed to conduct surveillance on the citizens of its partners and then share the results.[2] The NSA could not legally tap the phone of an American citizen without a warrant, but GCHQ could listen to that same phone call as it crossed the Atlantic. GCHQ could not surveil a British journalist without legal authority, but the NSA could collect that journalist's communications as they passed through American infrastructure. The intelligence was then shared across the alliance through secure channels. The legal restrictions of each nation were honored in the narrowest technical sense while being violated in every practical one.

For decades, this arrangement operated in near-total secrecy. The Australian government did not officially acknowledge its membership in the Five Eyes alliance until 2005.[2] Most citizens of the five member nations had never heard the term. The intelligence agencies preferred it that way. Secrecy was not a side effect of the alliance. It was a design feature.

The physical infrastructure of Five Eyes stretched across the planet. Listening stations dotted the landscapes of all five nations and their overseas territories. Satellite interception facilities tracked communications bounced off orbital relays. Undersea cable taps captured data flowing through the fiber optic lines that carry more than ninety-five percent of the world's international internet traffic. The system was called ECHELON, and for years its existence was treated as a conspiracy theory.[3]

ECHELON was a signals intelligence collection network capable of intercepting telephone calls, fax transmissions, and electronic mail on a global scale. The system used a network of ground stations positioned to capture satellite communications, along with taps on major telecommunications cables. Intercepted communications were filtered through a system of keyword dictionaries maintained by each member agency. When a communication matched certain criteria, it was flagged for human review.[3]

Governments denied ECHELON's existence for years. Journalists and civil liberties organizations who raised alarms about the system were dismissed as paranoid. The very idea that five allied governments had built a global communications interception network sounded too vast to be real. Then in 2001, the European Parliament published the results of a formal investigation that had taken more than a year to complete. The report confirmed that ECHELON was real, that it was operated by the Five Eyes alliance, and that it was capable of intercepting private and commercial communications on a massive scale.[3] The investigators concluded that the system had been used not only for national security purposes but also for economic espionage, giving member nations advantages in trade negotiations and competitive intelligence. European companies had reportedly lost contracts to American competitors who appeared to have advance knowledge of their bids.

The revelation should have been a watershed moment. A secret alliance of five governments had built a surveillance infrastructure that could monitor the communications of every person on the planet. They had done so without public debate, without legislative authorization in most cases, and without any meaningful mechanism for citizens to know whether their conversations had been intercepted.

The reaction was muted. The report was published in July 2001. Two months later, the attacks of September 11 redirected public attention toward security rather than privacy. In the calculus of a frightened public, the possibility that intelligence agencies were reading emails seemed less urgent than the certainty that nineteen men had turned commercial aircraft into weapons. The appetite for questioning surveillance evaporated almost overnight. The Five Eyes alliance did not contract in the wake of the European Parliament's findings. It expanded.

The British program known as Tempora illustrates the scale of that expansion. Operated by GCHQ, Tempora tapped directly into the undersea fiber optic cables that carry internet traffic into and out of the United Kingdom.[2] These cables are the physical arteries of the global internet. Nearly all communications between Europe and North America pass through them. GCHQ placed interception equipment at the points where these cables made landfall on British soil. The equipment copied the data flowing through the cables in real time. Content was stored for three days. Metadata, the records of who communicated with whom and when, was stored for thirty days.

The Five Eyes alliance collectively processes billions of communications each day.[2] The system operates across every time zone. When it is night in Maryland, it is morning in Cheltenham. When analysts at the NSA finish their shifts, their counterparts at GCHQ and ASD are already at their desks. Like Argus, some eyes are always open.

Five governments built a system to observe the entire world while remaining invisible themselves. They justified the secrecy as necessary for national security. They justified the surveillance as necessary for public safety. The logic was circular but effective. Trust us because you cannot see us, and you cannot see us because the work we do requires your trust.

Mark Klein spent twenty-two years working as a technician for AT&T. He maintained the switching equipment that routes telephone calls and internet traffic across the company's vast network. The work was technical and unglamorous. He kept the infrastructure running. He was the kind of employee that large corporations depend on but rarely notice.

In 2003, Klein was working at AT&T's Folsom Street facility in San Francisco when he observed something unusual. The company was constructing a secret room on the sixth floor of the building at 611 Folsom Street. The room was designated 641A.[4] Klein watched as workers from the NSA supervised its construction. Access to the room was restricted to personnel with NSA security clearances. AT&T employees without those clearances, including Klein himself, were not permitted inside.

Klein was a technician. He understood wiring diagrams and network architecture. He studied the documentation that crossed his desk and began to piece together what Room 641A was built to do. The room contained a fiber optic splitter, a device that copies light signals passing through a cable by creating a duplicate stream.[4] The splitter was connected to the building's main fiber optic lines, the cables that carried AT&T's internet traffic for a significant portion of the Western United States. Every piece of data that flowed through those cables was being duplicated. One copy continued on its intended path to its intended recipient. The second copy was diverted into Room 641A.

The traffic included emails, web browsing records, and internet phone calls. It included communications between Americans located within the United States.[4] This was not foreign intelligence collection directed at overseas targets. This was the wholesale copying of domestic internet traffic. A fiber optic splitter does not discriminate. It does not read the content that passes through it and select only the communications of foreign agents or suspected terrorists. It copies everything. Every email sent by an AT&T customer in the region, every website visited, every file transferred passed through the splitter and into the secret room where the NSA's equipment waited. The love letter and the terrorist communique received identical treatment. The medical consultation and the international conspiracy traveled the same path into the same room.

Klein retired from AT&T in 2004. He spent the next two years wrestling with what he had witnessed. He was not a political activist or a professional whistleblower. He was a technician who had seen something that troubled him and could not set it aside. He collected technical documents, wiring diagrams, and photographs. He prepared a sworn declaration describing the installation in precise detail. In 2006, he brought his evidence to the public.[4]

The Electronic Frontier Foundation used Klein's testimony and documentation as the foundation for a landmark lawsuit.[5] The case was called Hepting v. AT&T, and it alleged that AT&T had collaborated with the NSA to conduct an illegal and unconstitutional program of dragnet surveillance against the company's own customers.[5] Klein's technical diagrams showed exactly how the system worked. The fiber optic splitter was not a targeted surveillance tool. It was a net. It captured everything.

Klein also documented evidence suggesting that Room 641A was not unique. Similar installations existed at other AT&T facilities across the country, including locations in Seattle, San Jose, Los Angeles, and San Diego.[4] The San Francisco room was one node in a larger network. The

NSA had not placed a single wiretap on a single suspect's communications. It had wired itself into the backbone of the American internet.

The government's response followed a pattern that would become familiar. Officials did not deny that Room 641A existed. They did not claim that Klein's technical evidence was inaccurate. Instead, the Department of Justice intervened in the lawsuit by invoking the state secrets privilege, arguing that the case could not proceed because any litigation would require the disclosure of information that would harm national security.[5]

The legal battle continued for two years. Then in 2008, Congress passed the FISA Amendments Act.[6] The law contained a provision that retroactively granted legal immunity to telecommunications companies that had cooperated with government surveillance programs. AT&T and other companies that had opened their networks to the NSA could no longer be sued for their participation. The Hepting case was effectively killed.

The sequence deserves to be stated plainly. A technician discovered that the government was conducting mass surveillance of American citizens through secret rooms installed inside private telecommunications infrastructure. He gathered evidence and brought it to public attention. The government's response was not to dismantle the surveillance apparatus. The response was not to prosecute the agencies or officials responsible. The response was to pass a law that made the surveillance legal after the fact and shielded the participating companies from accountability.[5],[6] The government's answer was not to remove the ring. It was to legalize wearing it.

The Foreign Intelligence Surveillance Court was born from an earlier moment of reckoning. In the mid-1970s, a Senate committee chaired by Senator Frank Church of Idaho conducted a sweeping investigation into the intelligence community's domestic activities.[7] The Church Committee uncovered abuses that shocked the nation. The FBI had run a program called COINTELPRO that systematically targeted civil rights leaders, antiwar activists, and political dissidents for surveillance, harassment, and disruption. The NSA had operated a program called SHAMROCK that collected the contents of international telegrams sent by American citizens. The CIA had opened the mail of American citizens and maintained files on more than one and a half million people.[7]

The revelations led to a bipartisan consensus that the intelligence agencies needed judicial oversight. The solution was the Foreign Intelligence Surveillance Act of 1978, which created a specialized court to review government applications for surveillance warrants in national security cases.[7] The court was known formally as the Foreign Intelligence Surveillance Court, or FISC. Its informal name was the FISA Court.

The court was designed as a safeguard. When the government wanted to conduct surveillance on a person for intelligence purposes, it would present its case to a FISA Court judge. The judge would review the application, assess whether the government had met the legal standard, and approve or deny the request. The process was supposed to function as a check on executive power, a judicial hand on the brake that would prevent the intelligence agencies from returning to the unchecked surveillance practices that the Church Committee had exposed.

In practice, the court became something quite different. Its proceedings were classified. Its opinions were classified. Its docket was classified. The courtroom itself was a secured facility inside the Department of Justice building in Washington, shielded against electronic eavesdropping. When the government appeared before the court to request a surveillance warrant, no opposing counsel was present.[7] There was no defense attorney to challenge the government's claims. There was no offer of an alternative interpretation of the law. There was no representative of the person about to be surveilled. There was only the government, presenting its case to a judge in a sealed courtroom, and the judge, ruling on that case in secret.

The approval statistics tell a story that numbers rarely tell so clearly. Between 1979 and 2019, the FISA Court received 33,942 applications for surveillance warrants. It denied twelve.[7] The approval rate was 99.97 percent. In four decades of operation, the court that was created to serve as a check on surveillance power rejected government requests at a rate that rounds to zero.

Defenders of the court argued that the high approval rate reflected careful pre-submission screening by the Department of Justice, which weeded out weak applications before they reached the judges. That explanation contains a kernel of truth. But it also misses a larger reality. A court that approves virtually every request placed before it is not functioning as a meaningful check on power. It is functioning as a rubber stamp that creates the institutional appearance of oversight without its practical substance.

The court did more than approve individual warrants. Over the years, its secret rulings effectively rewrote surveillance law.[7] The most consequential example involved Section 215 of the Patriot Act, a provision that authorized the government to obtain business records relevant to an intelligence investigation. The plain language of the statute seemed to contemplate targeted requests for specific records. The FISA Court interpreted Section 215 differently. In a series of classified opinions, the court ruled that "relevant" could mean "all." Under this interpretation, the NSA was authorized to collect the telephone metadata of every American, because the entire database of records might contain information relevant to some future investigation.

This interpretation would have been controversial if it had been subject to public debate. It was not subject to public debate because the public did not know it existed. The court's opinions were classified. The legal reasoning that authorized mass metadata collection remained hidden for years, known only to the judges who wrote it, the government lawyers who argued for it, and the intelligence officials who operated under it.

The paradox completed itself with a kind of institutional elegance. A court created to watch the watchers had become invisible itself. It operated in darkness, issuing rulings that reshaped the relationship between citizens and their government without the knowledge or consent of either the citizens or their elected representatives. When Edward Snowden's disclosures forced some of these opinions into public view, they revealed something even more troubling. The court had been misled by the very agencies it was supposed to oversee.[7] Intelligence officials had provided inaccurate information about the scope of their collection programs. The court that was supposed to be the last line of accountability had been fed incomplete facts by the institutions it was supposed to hold accountable.

Quis custodiet ipsos custodes? The Roman poet Juvenal posed the question in his sixth Satire nearly two thousand years ago.[10] Who watches the watchers? The phrase echoes Plato's own concern about his guardian class. Both men understood that the most dangerous concentration of power occurs when those charged with oversight are themselves exempt from it. Both men understood that the question has no comfortable answer. In the case of the FISA Court, the answer was stark. The watchers watched themselves. The result was predictable.

On a Sunday in July 2021, a consortium of seventeen news organizations published the results of an investigation that had been months in the making. The project was coordinated by Forbidden Stories, a Paris-based journalism nonprofit, in partnership with Amnesty International's Security Lab. The investigation was called the Pegasus Project, and its findings reached into the governments of more than forty countries.[8]

At the center of the story was a company called NSO Group, based in Herzliya, Israel. NSO had developed a piece of software called Pegasus.[8] The name was apt. In Greek mythology, Pegasus was the winged horse that could go anywhere, reaching places no ordinary creature could access. The software that bore its name could do something comparable.

Pegasus was spyware designed to take complete control of a smartphone. Once installed on a target's device, Pegasus could read every text message and email on the phone. It could access every photograph. It could activate the microphone and listen to conversations happening in the room. It could turn on the camera and watch the phone's owner without their knowledge. It could track the device's location in real time, mapping every movement of the person who carried it.[8]

The technical sophistication was remarkable. Early versions of Pegasus required the target to click a malicious link sent via text message. A government agent would craft a convincing message, perhaps disguised as a shipping notification or a news alert, and embed the link inside it. When the target clicked, the software installed itself in seconds. Later versions eliminated even that requirement. NSO developed what security researchers call "zero-click" exploits, methods of installing Pegasus on a phone without any action from the target.[8] A missed call on WhatsApp was sufficient. An iMessage delivered to the phone could install the software before the owner ever opened it. The target did not need to click anything, open anything, or even touch the phone. The infection was silent and invisible. Once inside the device, Pegasus erased its own tracks so effectively that most targets never learned they had been compromised.

NSO Group marketed Pegasus exclusively to governments.[8] The company's stated purpose was counter-terrorism and law enforcement. Pegasus was supposed to help governments track terrorists, drug traffickers, and child predators. NSO maintained that it sold the software only to vetted government clients and that it investigated any reports of misuse.

The Pegasus Project revealed the gap between those claims and reality. The investigation was built on a leaked database containing more than fifty thousand phone numbers.[8] These numbers had been selected for potential surveillance by NSO's government clients. Forensic analysis of phones belonging to people on the list confirmed that many had been infected with Pegasus or had been targeted for infection.

The targets were not terrorists. They were not drug traffickers. The list included journalists from major news organizations across the globe, including reporters at The Washington Post, Le Monde, The Guardian, and Al Jazeera.[8] It included human rights activists who had challenged their governments' records on detention, torture, and political repression. It included opposition politicians who posed electoral threats to ruling parties. It included lawyers representing clients the state found inconvenient. It included at least one sitting head of state, French President Emmanuel Macron, whose phone number appeared among those selected for potential targeting by Moroccan intelligence.[8]

The countries implicated in the investigation read like a catalog of authoritarian and semi-authoritarian governance. Saudi Arabia was a major user of Pegasus.[8] Among the numbers associated with Saudi targeting were those belonging to associates of Jamal Khashoggi, the Washington Post columnist who was murdered by Saudi government agents inside the Saudi consulate in Istanbul in October 2018. Khashoggi had written columns critical of the Saudi crown prince. He had called for press freedom and political reform in the Arab world. His fiancée, Hatice Cengiz, was found to have been targeted with Pegasus in the days following his killing.[8] The targeting of Cengiz after the murder suggests that the surveillance apparatus did not shut down when its target was eliminated. It simply redirected itself toward those who remained.

The United Arab Emirates, Mexico, India, Hungary, and Morocco were among the other governments identified as NSO clients who had used the software against journalists and political opponents.[8] In Mexico, the numbers included those of journalists investigating drug cartel violence and government corruption. In Hungary, the targets included investigative reporters examining the business dealings of allies of Prime Minister Viktor Orban. The pattern repeated across borders. Wherever a government purchased Pegasus, it used the tool not against foreign enemies but against domestic critics.

The Pegasus story represents something that the history of surveillance had been building toward for decades. The ring had been privatized. NSO Group was a commercial enterprise. It employed software engineers who developed the spyware, salespeople who pitched it to government clients, and executives who collected the revenue. The company sold invisibility as a product. Governments purchased the ability to see into the most private spaces of their citizens' lives while remaining entirely unseen themselves.

No government using Pegasus disclosed its use to the public. No target was informed that their phone had been compromised. The entire system operated in darkness, visible only when security researchers or investigative journalists managed to detect the spyware's digital fingerprints on an infected device.[8]

The Pegasus revelations intersected with a story told earlier in this book. Daphne Caruana Galizia, the Maltese journalist whose investigation into government corruption was examined in Chapter 7, worked in an environment where Pegasus was reportedly deployed.[8] Malta was among the countries identified as a potential NSO client during the period of her reporting. Caruana Galizia spent years exposing what powerful people did when they believed no one was looking. She published her findings on a blog that drew more readers than any traditional media outlet in Malta. She tracked the money, named the names, and documented the connections that those in power wanted to keep hidden. The journalist who made the invisible visible may herself have been watched through the very phone she used to conduct her investigations. The irony is almost too precise to bear.

The technology that Hermes used to defeat Argus was narrative, the ability to control what the watcher paid attention to. The technology that governments used through Pegasus was the opposite. It was the ability to pay attention to everything while the target paid attention to nothing. The phone in a journalist's pocket became an open window through which the state could observe every professional contact, every confidential source, every draft of every story that might embarrass the powerful.

The pattern that emerges across these cases is not complicated. Intelligence agencies demand the right to see everything while insisting on the right to be seen by no one. They justify their invisibility as necessary for national security. They justify their surveillance as necessary for public safety. The argument has an internal consistency that makes it difficult to challenge on its own terms. Secrecy protects sources and methods. Surveillance prevents attacks. Both claims contain truth.

But the same logic, applied in the opposite direction, exposes the argument's fundamental asymmetry. If transparency is beneficial for citizens, then the question arises naturally... why is it not beneficial for the agencies that govern them? If citizens have nothing to hide, as surveillance advocates often insist, then the agencies should have nothing to hide either. If secrecy is genuinely necessary for the agencies to function, then perhaps secrecy is genuinely necessary for citizens to function as well. The argument cannot run in only one direction without revealing itself as an argument about power rather than principle.

The power asymmetry is enormous. The state possesses the largest ring. It commands the greatest resources to exploit that ring. It holds the strongest ability to prevent anyone from taking the ring away. When Mark Klein revealed Room 641A, Congress responded by immunizing the companies that helped build it.[6] When the European Parliament confirmed ECHELON, the revelations were overtaken by events within weeks.[3] When the FISA Court's secret interpretations of law were finally exposed, the intelligence community fought to reclassify them.[7] Every attempt to make the watchers visible was met with institutional resistance from the watchers themselves.

Plato understood this problem. His Republic is remembered primarily for its theory of justice and its allegory of the cave. But Plato devoted considerable attention to a question that modern democracies have largely failed to answer. What do you do with the guardians?[9]

Plato's guardians were the warrior class that protected his ideal city. They possessed physical strength, military training, and the authority to use force. They were, in effect, the security apparatus of the Republic. Plato recognized that giving one class of citizens the power to protect all others created an immediate danger. The guardians might decide to use their power not for the common good but for their own benefit. The protectors might become predators.

His solution was radical. The guardians would own no private property.[9] They would accumulate no wealth. They would maintain no private family ties. They would live in communal housing and eat at communal tables. Every incentive that might tempt them to abuse their power would be systematically removed. Plato did not trust his guardians to be virtuous on their own. He designed their entire way of life to make corruption as difficult as possible.

The reasoning was straightforward. The guardians had power that other citizens did not possess. That power created temptation. Temptation required constraint. The greater the power, the greater the constraint needed. Plato did not exempt his guardians from scrutiny because their work was important. He subjected them to the most intense scrutiny precisely because their work was important.

Modern intelligence agencies operate under none of Plato's safeguards. Their employees earn private salaries and accumulate private wealth. Their institutions develop institutional self-interest, competing for budget allocations and bureaucratic influence. Their leaders move between government service and private sector positions at defense contractors and consulting firms. The revolving door between the intelligence community and the industries that profit from surveillance creates financial incentives that Plato would have recognized immediately as incompatible with the guardian's role.

The agencies operate with enormous power, minimal oversight, and extensive capacity to resist the oversight that does exist. The FISA Court was supposed to constrain them, and it approved their requests at a rate of 99.97 percent.[7] Congressional intelligence committees were supposed to oversee them, and committee members have repeatedly acknowledged that they lack the technical expertise and the access to classified information necessary to conduct meaningful review.[9]

Inspectors general were supposed to audit them, and their reports are routinely classified.

The state's ring is the most dangerous ring because it is the hardest to take away. A corporation can be regulated, broken up, or bankrupted. A criminal can be prosecuted, imprisoned, or deterred. An anonymous internet troll can be unmasked, banned, or ignored. But a government that has built its surveillance apparatus into the physical infrastructure of global communications, that has established legal frameworks to authorize its own activities in secret, that has constructed courts and committees that provide the appearance of oversight without its substance... that government has achieved something that Gyges himself never managed. It has made the ring permanent.

Hermes killed Argus with patience and narrative, with the power to control what the watcher noticed. But the hundred-eyed giant keeps regenerating. In Hera's version of the myth, she honored the dead Argus by placing his hundred eyes on the tail of the peacock. The eyes could no longer watch, but they remained on permanent display as a reminder that Hera was always looking, even though dead instruments. The message was clear, the destruction of the watcher does not end the watching.

The state's ring carries the same message. Expose one program, and another takes its place. Shut down one secret room, and the infrastructure moves to a different building. Force one court to acknowledge the scope of surveillance, and the legal framework adapts to accommodate it. It was built to be permanent, to survive exposure, to regenerate after every revelation.

The hundred eyes remain open. Some of them watch the citizens they were built to protect. Some of them watch the journalists who ask uncomfortable questions. Some of them watch the activists who challenge the powerful. Some of them watch the politicians who threaten to reform the system. The eyes take shifts among themselves, and they never all close at once.

Who watches the watchers? The watchers are still deciding whether anyone needs to...

References

[1] Ovid, *Metamorphoses*, Book I (Argus Panoptes myth).

[2] Aid, Matthew M. *The Secret Sentry: The Untold History of the National Security Agency*. Bloomsbury Press, 2009.

[3] European Parliament. "On the Existence of a Global System for the Interception of Private and Commercial Communications (ECHELON Interception System)." Report, July 2001.

[4] Klein, Mark. *Wiring Up the Big Brother Machine... and Fighting It*. BookSurge Publishing, 2009.

[5] *Hepting v. AT&T*, Case No. C-06-0672-JCS (N.D. Cal.). EFF case documents.

[6] FISA Amendments Act of 2008, Pub. L. 110-261.

[7] Office of the Director of National Intelligence. FISA Court annual reports to Congress; Bamford, James. *The Shadow Factory: The NSA from 9/11 to the Eavesdropping on America*. Anchor Books, 2008.

[8] Forbidden Stories and Amnesty International. "The Pegasus Project." July 2021.

[9] Priest, Dana, and William M. Arkin. *Top Secret America: The Rise of the New American Security State*. Little, Brown and Company, 2011; Plato, *Republic*, Books III and IV.

[10] Juvenal. *Satires*. VI.347-348 ("Quis custodiet ipsos custodes?").

Chapter Twelve - The Glass Prison

In 1791, an English philosopher sat at his desk in London and designed the perfect prison. Jeremy Bentham was sixty-three years old. He had spent decades thinking about law, punishment, and the problem of human behavior. He believed that people act rationally in pursuit of pleasure and avoidance of pain. He believed that crime was a calculation, not a moral failing. Most importantly, he believed that a properly designed institution could change the calculation itself, making obedience more attractive than transgression without ever laying a hand on the prisoner.

The design he produced was elegant in its simplicity. Bentham called it the Panopticon, from the Greek for "all-seeing." The building was circular. Individual cells lined the outer ring, each one open to the interior and backlit by a window on the exterior wall. At the center of the circle stood a tower. Inside the tower sat a single watchman. From his position the watchman could see into every cell. From their positions the prisoners could see the tower but never the watchman inside it. They could never know at any given moment whether they were being observed. They could only know that they might be.[1]

The architecture encoded a profound asymmetry. Light flooded through the exterior windows, illuminating each prisoner in silhouette. The tower remained shrouded in shadow. A system of blinds and internal partitions ensured that no inmate could determine whether the watchman's gaze was directed at his cell or at any other. The prisoner was perpetually visible. The watchman was perpetually invisible. It was the ring of Gyges turned inside out and applied to the institution rather than the individual.

Bentham understood something important about the relationship between visibility and behavior. He grasped that actual surveillance was expensive, impractical, and often unnecessary. What mattered was not the watching itself but the possibility of being watched. If a prisoner internalized that possibility, if the awareness of the gaze became a permanent companion, then physical punishment became redundant. The prisoner would regulate his own behavior. He would become his own jailer.

This was not a cynical insight. Bentham was sincere. He was a reformer who detested the squalid cruelty of Georgian prisons, the flogging, the filth and the disease that festered in Newgate and the hulks moored in the Thames. He saw the Panopticon as humane. A prison without whips, a prison without chains. The architectural equivalent of a conscience, installed from the outside. He spent twenty years trying to convince the British government to build it. He lobbied Parliament. He purchased land in Tothill Fields. He sank his personal fortune into the project, eventually losing the equivalent of several million dollars in modern currency. The government never built it. Committees debated. Bureaucrats delayed. Political allies lost interest or lost office. Bentham died in 1832 believing his greatest idea had been wasted.[1]

He was wrong about the waste but not in the way he hoped. The Panopticon as a physical building was never constructed to his specifications. Elements of his design influenced prison architecture for the next two centuries. The Eastern State Penitentiary in Philadelphia adopted the radial layout. Pentonville Prison in London borrowed the principle of cellular isolation under centralized observation. Stateville Correctional Center in Illinois built a circular cellhouse that came closer to Bentham's original vision than any other structure in the world. But the Panopticon's true legacy was not architectural. It was philosophical. Bentham had articulated a principle that would come to define modern institutional power. The principle that the perception of being watched is itself a form of control.

Nearly two centuries later, a French philosopher named Michel Foucault read Bentham's writings and saw something the Englishman had not fully intended. In his 1975 book *Discipline and Punish*, Foucault argued that the Panopticon was not merely a prison design. It was a diagram of power itself.[2]

Foucault observed that the logic of the Panopticon had quietly spread far beyond prison walls. Schools arranged students in rows where teachers could observe them. Hospitals organized patients in wards where nurses could monitor them. Factories positioned workers along assembly lines where supervisors could watch them. Bureaucracies created hierarchies of observation where each level was visible to the one above it. In every case the same principle applied. Those being observed modified their behavior because they could never be certain they were not being watched.[2]

The genius of this arrangement was its efficiency. A single watchman in the tower could control hundreds of cells. A single supervisor on a factory floor could discipline dozens of workers. Power did not need to be exercised constantly. It only needed to be potentially present. The threat of observation was enough. Foucault called this the "automatic functioning of power," a system so well designed that it operated even when no one was actively wielding it.[2]

But Foucault pushed the analysis further. He argued that the Panopticon did not merely control behavior. It shaped identity. When people live under the permanent possibility of being watched, they do not simply choose to behave differently. They begin to think differently. They internalize the watching eye. They develop what Foucault called a "state of conscious and permanent visibility" that becomes part of who they are.[2] The prisoner does not merely obey the rules when the watchman looks his way. The prisoner becomes someone who obeys the rules, not out of genuine conviction but because the gaze has colonized his inner life.

Foucault identified three mechanisms through which this colonization proceeds. The first is hierarchical observation. The organization of space so that those in authority can see those beneath them. The second is normalizing judgment. The establishment of standards against which every individual is measured and found adequate or deficient. The third is the examination. The ritual in which the individual is made visible, documented, and classified.[2] Schools use all three when they arrange students in rows, grade them on curves, and subject them to standardized tests. Hospitals use all three when they organize patients in wards, measure their symptoms against diagnostic criteria, and record their conditions in charts. Prisons use all three by design. The modern workplace uses all three with increasing technological sophistication, as this chapter will demonstrate.

This insight brings us directly back to Plato's cave and to the ring of Gyges. If Glaucon's argument in *The Republic* is that invisibility unleashes the self, then Foucault's argument is that total visibility destroys it. The ring of Gyges removes the watcher and reveals what a person truly is. The Panopticon installs a permanent internal watcher and replaces what a person truly is with what the institution demands. One is a test of character. The other is a replacement of character.

Bentham believed he was building a machine for moral improvement. Foucault saw that he had built a machine for moral annihilation. The question is not whether Foucault was right in theory. The question is whether the evidence supports him in practice. In the twenty-first century, several governments have decided to test the hypothesis of total visibility at a scale Bentham could never have imagined. The results are now available for examination.

In 2014 the State Council of the People's Republic of China published a document titled "Planning Outline for the Construction of a Social Credit System." The language was bureaucratic. The ambition was extraordinary. The Chinese government proposed to create a comprehensive system for evaluating the trustworthiness of every citizen, corporation, and government official in a country of 1.4 billion people. The system would track behavior across multiple domains and assign consequences accordingly. Those who scored well would be rewarded with preferential access to services. Those who scored poorly would be punished with restrictions that cascaded across every dimension of modern life.[3]

The idea had been circulating in Chinese policy circles since at least 2007. Its origins lay in a practical problem. China's rapid economic transformation had created a society where traditional mechanisms of trust had broken down. People moved from villages where everyone knew them to cities where no one did. Contracts were broken without consequence. Fraud was rampant. The government struggled to enforce court judgments. In this context, the social credit system was presented not as an instrument of political control but as a tool for building trust in a society that lacked it.[3]

The Western media frequently describes the social credit system as a single unified score, a kind of national grade assigned to every Chinese citizen by a central algorithm. This is a common misconception. The reality is more fragmented and in some ways more unsettling. What exists is a patchwork of systems. Municipal governments run local credit programs with different criteria and different scoring methods. Commercial companies like Sesame Credit, operated by Alibaba's Ant

Financial, assign their own scores based on consumer behavior. The central government maintains national blacklists for specific offenses. These systems overlap, interact, and sometimes contradict each other. There is no single number that defines a Chinese citizen's social worth. There are many numbers, maintained by many authorities, using many criteria that are not always transparent to the people being scored.[3]

What these systems share is a common architecture of consequences. The most visible tool is the blacklist system. By 2019 the Supreme People's Court of China reported that individuals on the courts' blacklist for failing to comply with legal judgments had been blocked from purchasing airplane tickets over 26.4 million times. They had been blocked from purchasing high-speed train tickets over 5.96 million times.[4] These numbers represent instances of denial, not unique individuals. But even adjusted for repetition the scale is staggering. Millions of Chinese citizens found themselves unable to travel freely because a database had flagged them as untrustworthy.

The consequences extend beyond travel. Some localities display the photographs and names of blacklisted individuals on public billboards and smartphone apps. Others restrict access to schools, hotels, and government services. Finding employment becomes difficult when a potential employer can check a credit database before making a hiring decision. The system creates a cascading series of penalties that can fundamentally alter a person's ability to participate in society. A single mark on a single record can ripple outward through housing, education, employment, and social life until the affected individual finds every door in the country closed.

Consider the case of Liu Hu. Liu was a journalist who had reported on corruption among Chinese officials. He was sued for defamation and lost the case. After the judgment was entered against him, he was placed on the Supreme Court's blacklist for failing to pay the fine and issue a public apology. He subsequently discovered that he could not book airline tickets. He could not purchase high-speed train tickets. He could not obtain bank loans. He could not buy property. A single legal judgment had effectively made him a second-class citizen, frozen out of the infrastructure of modern Chinese life.[9]

Liu Hu is a journalist, and his case attracted international attention. But the system operates most powerfully in its anonymous application to ordinary people. In several Chinese cities, facial recognition cameras at intersections capture the faces of jaywalkers. The photographs are displayed on large public screens near the intersection, sometimes within seconds of the violation. The person's name and partial identification number appear alongside their image. The purpose is explicit. Shame is the mechanism of behavioral correction. The camera is the watchman. The billboard is the tower. The pedestrian crossing is the cell.[9]

This is the vision of the social credit system in its purest form. A society in which no one possesses the ring of Gyges. Every action is visible, every action is recorded, and every action is scored. The philosopher's thought experiment has been made real through the infrastructure of surveillance technology. The question is whether the vision works. Does total visibility produce the virtue that Bentham promised and that Glaucon doubted?

The evidence suggests something more complicated. Researchers studying the social credit system have documented widespread anxiety and self-censorship among Chinese citizens. People modify their behavior not because they have internalized new values but because they fear the consequences of a low score. They perform trustworthiness for the cameras. They curate their public behavior the way a social media user curates a profile. Behind the performance, hidden from the cameras and databases, life continues in all its messy human complexity.[3]

The system has also produced a thriving economy of evasion. Citizens learn which behaviors are tracked and which are not. They learn to game the score, engaging in visible acts of compliance while concealing the behaviors that actually matter to them. They become performers on a stage built by the state, delivering the lines the audience demands while keeping their true selves hidden in the wings. Some purchase goods from blacklisted individuals through intermediaries. Others use unofficial communication channels that the monitoring systems cannot reach. The ring of Gyges has not been destroyed. It has simply been driven underground, where it operates with the added urgency of people who know what happens when the ring fails. Foucault predicted exactly this outcome. The Panopticon does not produce good people. It produces people who are good at appearing good. The distinction matters enormously.

The logic of total visibility reaches its most extreme expression in the Xinjiang Uyghur Autonomous Region of northwestern China. What has been constructed there represents the most intensive surveillance system ever built, a laboratory for the complete destruction of the ring of Gyges applied to an entire population of approximately twelve million people.

The physical infrastructure is overwhelming in its scope. Facial recognition cameras stand at every major intersection, every building entrance, and every checkpoint. The density of surveillance coverage leaves virtually no public space unmonitored. Residents must pass through checkpoints with identification cards to move between neighborhoods. Their vehicles are tracked by GPS. Their phones are scanned for content the government deems suspicious.[5]

The digital infrastructure is equally comprehensive. Residents are required to install mandatory applications on their smartphones that scan the device for files, images, and communications flagged as extremist or separatist. The government has collected DNA samples, iris scans, and voice recordings from residents during compulsory health screenings. All of this biometric data flows into databases that can identify any individual in any context. A person's face, voice, genetic code, and patterns of movement through physical and digital space have all been catalogued and made available to the security services at the touch of a screen.[5]

Tying these systems together is the Integrated Joint Operations Platform, known by its acronym IJOP. This artificial intelligence system aggregates data from surveillance cameras, wifi routers, phone scanners, checkpoints, and human informants. It processes this ocean of data and flags individuals for further investigation or detention. The criteria for flagging are broad enough to encompass almost any behavior. Praying more frequently than usual can trigger an alert. Using a VPN, communicating with relatives who live abroad, purchasing an unusual amount of fuel, or simply owning too many books can generate the same result. The system produces lists of names and sends them to local security officials, who are expected to act on them promptly.[5]

Between 2017 and 2019, the IJOP and the broader security apparatus fed hundreds of thousands of Uyghurs and other Turkic Muslims into a network of detention facilities that the Chinese government describes as "vocational education and training centers." Independent researchers estimate that between one million and 1.8 million people were detained in these facilities at the peak of the campaign.[5] The numbers are difficult to verify precisely because the system that monitors every action of Xinjiang's residents operates behind its own wall of secrecy. The watchman sees all. The watchman is seen by no one. Satellite imagery analyzed by the Australian Strategic Policy Institute revealed the construction of hundreds of new facilities across the region between 2017 and 2020, many of them surrounded by guard towers, razor wire, and high concrete walls that were difficult to reconcile with the claim that these were voluntary educational institutions.

The Chinese government insisted that these facilities were schools, offering job training and Chinese language instruction to willing students. Leaked internal documents told a different story. The "China Cables," a set of classified directives obtained by the International Consortium of Investigative Journalists in 2019, revealed that the facilities operated under strict security protocols. Detainees were forbidden from leaving. Their movements within the facilities were tightly controlled. They were subjected to compulsory political indoctrination sessions lasting hours each day. They were punished for speaking the Uyghur language. They were required to demonstrate gratitude to the Communist Party and to President Xi Jinping personally.[5]

In 2022 a massive trove of files hacked from police servers in Xinjiang became public. The "Xinjiang Police Files" contained thousands of photographs of detainees, internal police speeches, and operational documents. The photographs showed elderly men and women alongside teenagers, their faces blank with the particular emptiness of people who have been stripped of agency. Some of the detainees were as young as fifteen. Others were in their eighties. The documents revealed that guards were authorized to shoot detainees who attempted to escape. Internal speeches by senior officials made clear that the purpose of the facilities was not education but the systematic transformation of an entire population's religious and cultural identity.[6]

This is the logical endpoint of destroying the ring of Gyges entirely. When the state decides that no one should be invisible, when it builds the infrastructure to ensure that every prayer, every phone call, every footstep is monitored and recorded and analyzed, the result is not virtue. The result is not the moral improvement that Bentham envisioned. The result is a prison without walls, a society in which the entire population lives inside the Panopticon, watched by an eye that never sleeps and never blinks. Bentham imagined his prisoners would internalize the gaze and become self-governing. The Uyghurs internalized the gaze and became something else entirely. They became haunted.

The people of Xinjiang did not become better under total surveillance. They became terrified. Those who could, fled the region and then the country. Those who could not, retreated into a performance of compliance so total that it consumed their identity. They stopped praying or prayed only when ordered to during state-supervised sessions. They stopped speaking their language or spoke it only in whispers behind closed doors. They stopped being who they were because who they were had been classified as a threat to state security.

Bentham believed that if people internalized the gaze, they would self-regulate and physical punishment would become unnecessary. In Xinjiang the population internalized the gaze completely. Physical punishment proved entirely necessary anyway, because the behavior the state demanded was not mere lawfulness. It was the erasure of cultural identity. No amount of architectural cleverness or algorithmic sophistication could make that erasure voluntary. The Panopticon required the gulag after all.

There is a natural temptation at this point in the chapter to draw a firm line between the surveillance practices of authoritarian China and the freedoms of the democratic West. The temptation should be resisted. The Western world has constructed its own panoptic systems. They are softer in their application and less centralized in their design. They are no less pervasive in their reach.

Let's begin with credit scores. In the United States, three private companies dominate the credit reporting industry. Equifax, Experian, and TransUnion. These companies collect financial data on virtually every American adult. They aggregate that data into a three-digit score that determines whether a person can rent an apartment, buy a car, purchase a home, or in some cases get a job. The algorithms that generate these scores are proprietary. The data they rely on is frequently inaccurate. A 2013 Federal Trade Commission study found that one in four consumers identified errors on their credit reports that might affect their scores. Disputing those errors is a process so byzantine that it functions as its own form of punishment. Yet this score, calculated by private corporations using opaque methods, shapes the material conditions of hundreds of millions of lives.

The parallels to China's social credit system are uncomfortable precisely because they are imperfect. No one is displayed on a billboard for a low FICO score. No one is prevented from boarding a train. But the architecture is the same. A distant and invisible authority collects data about your behavior, processes that data through an algorithm you cannot examine, and produces a judgment that constrains your freedom. The panoptic principle operates with equal efficiency whether the watchman in the tower is a government bureau or a corporate server farm.

Background checks extend the surveillance further. Criminal records follow Americans for life in most states, regardless of rehabilitation or the passage of time. Employment history is aggregated and sold. Social media screening has become routine in hiring. A 2018 CareerBuilder survey found that seventy percent of employers reported checking candidates' social media profiles during the hiring process. Forty-three percent reported monitoring current employees' social media activity. The reasons ranged from checking for "inappropriate" content to evaluating "cultural fit," a phrase so vague that it can justify almost any hiring preference or any hiring exclusion.

Sex offender registries represent perhaps the most explicit American version of permanent visibility. In every state, individuals convicted of certain offenses are placed on publicly accessible databases that display their names, photographs, addresses, and offense details. In many states registration is required for life. The practical consequences mirror those of China's blacklist system with striking precision. Registrants face restrictions on where they can live, where they can work, and where they can go. Research has shown that the registries produce profound social isolation, homelessness, and unemployment among those listed on them. The stated purpose is public safety. The operational effect is the permanent destruction of the ring of Gyges for a designated class of citizens, who live under a gaze from which no rehabilitation, no passage of time, and no change of character can ever free them.

These systems differ from China's social credit architecture in important ways. They are decentralized rather than coordinated. They are maintained by private companies rather than a central government. They lack the explicit ideological purpose that drives the Chinese system. But these differences, real as they are, can obscure a deeper structural similarity. In every case an authority that the individual cannot see or challenge observes that individual's behavior, renders a judgment, and imposes consequences. In every case the individual is expected to modify behavior in response to the knowledge that the judgment is being rendered. In every case the relationship between the watcher and the watched is asymmetric. The watcher knows everything about the watched, while the watched knows almost nothing about the criteria by which they are being judged.

These systems share a common assumption with Bentham's original design. They assume that if you make behavior visible and attach consequences to that visibility, people will behave better. They assume that surveillance produces virtue.

The assumption was tested with unprecedented thoroughness during the COVID-19 pandemic, when millions of knowledge workers shifted to remote employment and their employers faced a sudden crisis of invisible labor. Workers were no longer visible in office buildings where managers could walk by their desks. They had vanished into their homes.

They had found a kind of ring of Gyges, and their employers were not comfortable with it.

The market for employee monitoring software exploded. Products like Hubstaff, Time Doctor, and Teramind offered employers the ability to track every keystroke their employees typed. These programs captured screenshots of employee monitors at random intervals, sometimes every five minutes, sometimes every thirty seconds. They tracked mouse movement to determine whether workers were actively engaged with their computers. Some offered webcam monitoring that could verify whether the employee was physically present at their desk. Teramind's marketing materials boasted of the ability to detect "insider threats" through behavioral analysis, flagging patterns of activity that the algorithm deemed suspicious. Time Doctor offered a feature that would display a pop-up notification asking "Are you still working?" if it detected a period of inactivity. The pop-up was the digital equivalent of the watchman tapping on the glass. The software industry had found a way to put every remote worker in a virtual Panopticon, their home offices transformed into cells with windows that opened onto the central tower of management oversight.

The growth of this industry was extraordinary. Research firms estimated that the global market for employee monitoring software grew by over sixty percent between 2019 and 2022. Millions of workers who had never experienced digital surveillance suddenly found their every mouse click logged and catalogued. Some adapted by installing programs that jiggled their mouse automatically to simulate activity. Others kept secondary devices running work applications while they stepped away. The surveillance produced its own counter-surveillance in an arms race that would have been darkly familiar to Foucault. The watchman in the tower has been replaced by an algorithm. The effect on the worker remains the same.

The documented consequences of workplace surveillance are consistent across industries and countries. A comprehensive review of electronic performance monitoring research published in the *Journal of Management* in 2020 found that monitoring increased compliance with specific measurable tasks. Workers who knew they were being watched met their numerical targets more reliably.[8] This finding supports Bentham's hypothesis at its most basic level. The presence of the watcher does change behavior.

But the same research found that monitoring decreased creativity, innovation, and intrinsic motivation. Workers who felt constantly observed reported higher levels of stress, anxiety, and emotional exhaustion. They were less likely to take initiative, less likely to experiment with new approaches, and less likely to report genuine engagement with their work.[8] They performed the observable behaviors the system demanded. They ceased to perform the unobservable behaviors that organizations actually need. The spirit of collaboration, the spark of invention, the willingness to take risks in pursuit of something better... these qualities withered under the watchman's gaze.

The evidence against the hypothesis of total visibility has been accumulating for decades. It is now substantial enough to state the conclusion plainly. Total surveillance does not produce virtue. It produces compliance. The distinction between these two things is the central finding of this chapter.

Consider the most rigorous studies available. The United Kingdom has one of the highest concentrations of surveillance cameras in the world. Estimates place the number of CCTV cameras across the country at approximately six million, roughly one camera for every eleven citizens. The British experience provides a natural experiment in the effects of pervasive public surveillance on behavior. In 2009 criminologists Brandon Welsh and David Farrington published an updated systematic review and meta-analysis of studies examining the relationship between CCTV and crime. They reviewed forty-four studies conducted across multiple countries. Their findings were instructive but modest.[7]

CCTV showed a statistically significant effect on crime reduction in car parks, where cameras could observe a contained area with limited access points. The effect amounted to a fifty-one percent reduction in vehicle crimes in those specific enclosed spaces.[7] In city centers, public housing projects, and public transportation systems, the effect on crime was not statistically significant. CCTV had essentially no measurable impact on violent crime in any setting. Six million cameras across an entire country produced modest reductions in car theft from parking garages. The Panopticon, it turned out, was effective only in spaces that already resembled the Panopticon... enclosed, limited, and focused on property rather than persons.

Body cameras, which we will examine in the next chapter, represent one attempt to make the watchers visible. Early studies produced dramatic results, but the pattern they revealed was consistent with the broader surveillance literature. Initial compliance effects that faded as the technology became routine.

The pattern that emerges from these studies is consistent. Surveillance produces behavioral changes that are specific, temporary, and shallow. Cameras in parking garages reduce car theft because car theft in a parking garage is an observable behavior in a contained space. Scale the surveillance up, make it routine, make it permanent, and the effect dissipates. People adapt. They learn the blind spots. They perform for the cameras in the spaces the cameras cover and revert to their baseline behavior everywhere else.

The workplace monitoring literature confirms this pattern with particular clarity. The 2020 review by Ravid and colleagues found that electronic performance monitoring reliably increases what the authors call "task performance," meaning compliance with specific measurable outputs. If you tell a worker you are counting their keystrokes, they will type more keystrokes. If you tell a warehouse worker you are timing their pick rate, they will pick faster.[8]

But the same review found that monitoring reduces what the researchers call "contextual performance" and "creative performance." These categories encompass behaviors that organizations depend on but cannot easily measure. Helping a colleague with a problem, suggesting a new approach to an old process, taking the initiative on a project that nobody assigned, persisting with a difficult problem after the obvious solutions have failed.[8] These behaviors require autonomy and a sense of ownership. They require the feeling that one is acting from internal motivation rather than external compulsion. Surveillance erodes all of these conditions. The watched worker becomes a compliant worker. The compliant worker ceases to be a creative one.

This finding illuminates something fundamental about the

relationship between visibility and virtue. Bentham assumed that self-regulation would be a positive development, that people who policed their own behavior would become better people. Foucault saw the flaw in this assumption. Self-regulation under surveillance is not the same thing as self-governance through conviction. The prisoner who behaves because he fears the watchman has not developed a conscience. He has developed a performance. Remove the watchman and the performance ends.

The ring of Gyges was supposed to be the test case for human nature. Plato's question was what people become when no one is watching. Bentham's implicit counterargument was that the question did not matter, because you could design a world in which someone was always watching. The data from two centuries of experimentation with that counterargument is now clear. You cannot build virtue through surveillance. You can build compliance, anxiety, and performance. You can build a workforce that meets its metrics while losing its soul. You can build a citizenry that obeys the law while abandoning genuine moral reasoning. You can build a prison that looks like a society and a society that functions like a prison. What you cannot build is the thing Bentham actually wanted, which was a population of people who do the right thing because they have come to understand that it is right.

Return for a moment to the circular building that Jeremy Bentham designed in 1791. Picture the cells arranged around the perimeter. Picture the tower at the center. Picture the watchman who may or may not be watching and the prisoners who can never know.

Now picture a variation. In this version the walls of the cells are transparent. The ceiling is transparent. The floor is transparent. Every prisoner can see every other prisoner at all times. Every movement, every expression, every private moment is exposed not just to the watchman in the tower but to every other inmate in the facility. This is the glass prison. This is the world that total surveillance creates.

In such a prison every inmate performs every action for an audience that includes every other inmate and the unseen authority above them all. There is no backstage. There is no moment of solitude in which a person can test their own convictions against their own desires. No one develops virtue in such a prison. The condition for virtue is choice, and genuine choice requires the possibility of acting without being observed. A person must be free to do the wrong thing in order for doing the right thing to carry moral weight. The prisoner in the glass cell cannot do the wrong thing. He cannot do anything at all without being observed. His compliance is not a moral achievement. It is a structural inevitability. He has not become good. He has become transparent, and transparency under compulsion is not goodness. It is the absence of the conditions under which goodness becomes possible.

The social credit system in China does not produce trustworthy citizens. It produces citizens who perform trustworthiness under conditions where the performance can be verified. The workplace monitoring software does not produce dedicated employees. It produces employees who generate the metrics that the software is designed to measure. The CCTV cameras that blanket British cities do not produce law-abiding communities. They produce communities where certain crimes migrate from surveilled spaces to unsurveilled ones. In every case the surface changes while the substance remains untouched.

Surveillance changes observable behavior in observable spaces for as long as the surveillance operates. It does not change the person. The person remains who they are, hidden behind the performance, waiting in the wings for the moment when the camera blinks or the algorithm looks away. The self that the Panopticon was supposed to improve has simply learned to hide more effectively. It has not become virtuous. It has become strategic.

Foucault was right, and his insight can be stated with a clarity he himself rarely achieved. The Panopticon does not produce good people. It produces people who are good at performing goodness. The watched person does not develop an internal moral compass. The watched person develops an external one, oriented always toward the watcher, calibrated always to the watcher's expectations. Remove the watcher and the compass spins freely, pointing nowhere at all.

This is the deepest failure of the anti-ring. If the ring of Gyges reveals who we truly are in the absence of observation, then the Panopticon reveals nothing at all. It replaces the self with a performance. It substitutes compliance for character. It answers Glaucon's challenge not by proving him wrong but by making his question irrelevant. In a world of total surveillance, it does not matter whether people are naturally good or naturally selfish. Neither quality has room to express itself. The only thing that expresses itself is the architecture of control.

The Panopticon makes its inmates invisible to themselves. That is the cruelest irony of the glass prison. The structure that was built to see everything ends up seeing nothing that matters.

References

[1] Bentham, Jeremy. *Panopticon; or, the Inspection-House*. 1791.

[2] Foucault, Michel. *Discipline and Punish: The Birth of the Prison*. Translated by Alan Sheridan. Vintage Books, 1977 (originally 1975).

[3] Creemers, Rogier. "China's Social Credit System: An Evolving Practice of Control." University of Leiden, 2018.

[4] Supreme People's Court of China. National Blacklist and Credit Enforcement Statistics, 2019.

[5] Zenz, Adrian. "Brainwashing, Police Guards, and Coercive Internment: Evidence from Chinese Government Documents about the Nature and Extent of Xinjiang's 'Vocational Training Internment Camps.'" *Journal of Political Risk* 7, no. 7 (2019).

[6] Xinjiang Police Files. Published by Adrian Zenz and the Victims of Communism Memorial Foundation, 2022.

[7] Welsh, Brandon C., and David P. Farrington. "Public Area CCTV and Crime Prevention: An Updated Systematic Review and Meta-Analysis." *Justice Quarterly* 26, no. 4 (2009): 716-745.

[8] Ravid, Daniel M., et al. "EPM 20/20: A Review, Framework, and Research Agenda for Electronic Performance Monitoring." *Journal of*

Management 46, no. 1 (2020): 100-126.

[9] Chin, Josh, and Liza Lin. *Surveillance State: Inside China's Quest to Launch a New Era of Social Control*. St. Martin's Press, 2022.

Chapter Thirteen - Everybody's Watching

The plumber heard it before he saw it. George Holliday was standing in his apartment in Lake View Terrace, a quiet neighborhood tucked into the northeastern corner of the San Fernando Valley. The date was March 3, 1991. The time was just after midnight. The sound coming through his window was the unmistakable crack of wood against flesh, repeated over and over, accompanied by shouting and the whine of sirens. Holliday had recently purchased a new Sony Handycam. He grabbed it off the shelf, stepped onto his balcony, and pressed record.

What the camera captured over the next eighty-one seconds would alter the relationship between citizens and police in America forever. On the ground below, a Black man named Rodney King lay on the pavement while a group of Los Angeles Police Department officers surrounded him. Four officers delivered the blows while more than a dozen others stood nearby and watched. The officers struck King fifty-six times with metal batons. They kicked him at least six times. King suffered skull fractures, broken bones in his face, and permanent brain damage.[1]

Before that night, police brutality in Los Angeles was an open secret discussed in whispers. Black and Latino residents of the city had reported excessive force for decades. The complaints followed a predictable pattern. A civilian would file a report. The officer's account would contradict it. Internal review boards would side with the officer. The case would disappear. The system relied on a simple asymmetry... the officer's word carried institutional weight, while the civilian's word carried none. It was the ring of Gyges operating through a badge. The officer's authority made his version of reality the only version that mattered, because nobody could prove otherwise.

Holliday's tape shattered that asymmetry overnight. He brought the footage to a local television station, KTLA, which broadcast it on the evening news. Within forty-eight hours the tape had been picked up by every major network in the country. Millions of Americans watched the beating play out on their television screens. For many white viewers, it was the first time they had ever witnessed the reality that Black communities had described for generations. The tape did not show a disputed encounter with two conflicting accounts. It showed a man on the ground being beaten by officers who faced no resistance. The footage was unambiguous.

The four officers were charged with assault and excessive use of force. Their trial was moved from Los Angeles to the predominantly white suburb of Simi Valley. On April 29, 1992, the jury acquitted all four officers on all charges. Los Angeles erupted. Over six days of rioting, sixty-three people were killed. More than two thousand were injured. Property damage exceeded one billion dollars. The tape had made the beating visible to the entire nation, but visibility alone had not produced justice.[1]

The acquittal revealed something important about the limits of watching. The tape could show what happened. It could not force a jury to care. It could not override the racial biases that allowed twelve people to watch a man being beaten and conclude that the officers had acted reasonably. The camera could strip away invisibility, but it could not strip away indifference. That distinction would prove essential in the decades to come.

Still, the Holliday tape established a precedent that would reshape American civic life. A single bystander with a camera could bypass every institutional barrier that had kept police violence hidden. The civilian did not need to file a complaint. He did not need to persuade an internal review board. He did not need the cooperation of the police department. All he needed was a lens and a record button. For the first time in history, ordinary citizens possessed a tool that could make the powerful visible against their will. The ring of authority, which had allowed officers to act without accountability for so long, had developed a crack. A plumber with a camcorder had put it there.

Sixteen years after George Holliday stood on his balcony, Steve Jobs walked onto a stage in San Francisco and introduced the iPhone. The device was marketed as a phone, a music player, and an internet communicator. Nobody described it as a surveillance tool. Nobody called it a weapon against the powerful. But that is precisely what it became.

The key transformation was not the camera itself. Camcorders had existed for years. The transformation was the elimination of intent. Before the smartphone, recording required a deliberate decision. A person had to own a camcorder, charge it, bring it to a specific location, and turn it on at the right moment. George Holliday's tape was a product of extraordinary luck. He happened to own a camera. He happened to hear the commotion. He happened to step outside at the right time. If any one of those conditions had been different, the beating of Rodney King would have remained invisible.

The smartphone collapsed all of those conditions into a single object that people carried everywhere. By 2015, more than two billion people worldwide owned smartphones with high-definition cameras.[2] The device was always charged, always present, always ready. Recording no longer required preparation. It required only the instinct to raise a hand and tap a screen. The age of accidental witnessing had arrived.

The consequences emerged in a cascade of recordings that exposed police violence with devastating clarity. On July 17, 2014, a twenty-two-year-old named Ramsey Orta stood on a sidewalk in the Tompkinsville neighborhood of Staten Island. He pointed his phone at a group of NYPD officers who had confronted his friend, a forty-three-year-old Black man named Eric Garner. Garner had been accused of selling loose cigarettes. Officer Daniel Pantaleo approached Garner from behind, wrapped his arm around Garner's neck, and pulled him to the ground. The chokehold was a technique explicitly banned by NYPD policy. On the video, Garner can be heard repeating the same words eleven times as the officers held him down, "I can't breathe." Eric Garner died on that sidewalk. The medical examiner ruled his death a homicide.[3]

The grand jury declined to indict Officer Pantaleo. The video had captured the killing in full. It had recorded the banned technique, the repeated pleas, the officers who stood by and did nothing. None of it was enough to produce a criminal charge. The pattern from the Rodney King case was repeating itself. Visibility did not guarantee accountability.

Less than a year later, on April 4, 2015, a different recording produced a different result. A barber named Feidin Santana was walking to work in North Charleston, South Carolina, when he saw a police officer and a Black man struggling on the ground. Santana pulled out his phone and began recording. On the video, Walter Scott breaks free from Officer Michael Slager and runs away. Scott is unarmed. He is running in the opposite direction. Slager draws his weapon and fires eight shots at Scott's back. Scott falls. He does not get up. On the video, Slager then walks to Scott's body and appears to drop an object near it. Slager had initially told investigators that Scott had taken his Taser and that he had feared for his life. The video proved both claims were false.[4]

Without Santana's recording, Slager's story would have become the official account. Scott's death would have been filed as a justified shooting. The investigation would have relied entirely on the officer's word. Santana's phone stripped away the lie. Slager was charged with murder. He later pleaded guilty to federal civil rights violations and was sentenced to twenty years in prison. The camera had done what decades of civilian complaints could not. It had made an officer's lie impossible to sustain.

Then came Minneapolis. On May 25, 2020, a seventeen-year-old named Darnella Frazier was walking to a corner store with her nine-year-old cousin. Outside the store, she saw a police officer kneeling on the neck of a Black man who was lying face down on the pavement. The man's name was George Floyd. The officer's name was Derek Chauvin. Frazier pulled out her phone and recorded.[5]

For nine minutes and twenty-nine seconds, Frazier held her phone steady while Floyd begged for his life. He called for his mother. He said he could not breathe. Bystanders pleaded with Chauvin to check Floyd's pulse. Chauvin did not move. He kept his knee on Floyd's neck even after Floyd fell silent, even after Floyd stopped moving, even after the paramedics arrived. Frazier recorded all of it.

The video went viral within hours. Protests erupted across the United States and then spread to countries around the world. Chauvin was arrested, charged, and convicted of second-degree murder, third-degree murder, and second-degree manslaughter. Frazier's video was the central piece of evidence at trial. Without it, the Minneapolis Police Department's initial press release would have stood as the official account. That press release described Floyd's death as a "medical incident" that occurred after officers responded to a forgery complaint. It made no mention of a knee on a neck. It made no mention of nine minutes.[5]

At the trial, Frazier testified about what it had cost her to stand there and record. She told the court that she stayed up at night and apologized to George Floyd. "I apologized to him for not doing more and not physically interacting and not saving his life," she said. She was seventeen years old when she witnessed a man die. She did not choose to become a witness. The moment chose her, as it had chosen Holliday and Orta and Santana. The camera does not simply record events. It conscripts the person holding it into a role that carries weight for the rest of their lives. Frazier won a special Pulitzer Prize citation in 2021 for her courage, but she also testified about anxiety, insomnia, and the inability to stop replaying what she had seen.[5]

The camera phone had proven itself as a tool of accountability. It could make police violence visible. It could contradict official narratives. It could produce convictions that would have been impossible without visual evidence. But the people who wielded it paid a price that the public rarely considered. Ramsey Orta, who filmed Eric Garner's death, was subsequently arrested multiple times on charges his supporters described as retaliatory. He spent years in prison. The camera could protect the public, but it could not always protect the person who held it.

The impulse to watch the watchers did not emerge spontaneously from the smartphone revolution. It had roots in organized movements that predated the iPhone by decades. In 1990, a group of activists in Berkeley, California, founded an organization called Copwatch. Their method was straightforward. Volunteers would patrol the streets carrying video cameras, documenting encounters between police officers and civilians. When an officer stopped someone, a Copwatch volunteer would stand at a legally permissible distance and record the interaction. The goal was deterrence through visibility. If officers knew they were being recorded, the theory went, they would be less likely to use excessive force.[6]

The concept spread to cities across the country. Similar groups emerged in New York, Phoenix, Los Angeles, and dozens of other communities. The model also crossed borders. Cop-watching movements appeared in Mexico, Brazil, the United Kingdom, and Hong Kong. The underlying principle was always the same... power operates differently when it knows it is being observed. Steve Mann, a researcher at the University of Toronto, coined a term for this practice. He called it "sousveillance," from the French word "sous," meaning "from below."

Traditional surveillance watches from above. The state monitors its citizens. Sousveillance inverts that relationship. The citizens monitor the state.[8]

Technology amplified the practice enormously. The American Civil Liberties Union developed a smartphone application called Mobile Justice that allowed users to record police encounters and automatically upload the footage to ACLU servers in real time. The design addressed a specific problem. Officers who were caught on camera sometimes confiscated or destroyed the recording device. Mobile Justice made confiscation irrelevant. By the time an officer could seize a phone, the footage was already stored on a remote server beyond the officer's reach.

The institutional response to citizen recording was body cameras. If civilians were going to record officers, then departments would record their own officers first and control the narrative. The idea gained momentum after a landmark study conducted in Rialto, California, in 2012. Researchers outfitted half of the department's officers with body cameras while the other half served as a control group. The results were striking. Officers wearing cameras showed a sixty percent reduction in use-of-force incidents. Complaints against officers wearing cameras dropped by eighty-eight percent. The study seemed to confirm what Bentham had theorized about the Panopticon two centuries earlier, visibility produces compliance.[3]

Police departments across the country adopted body cameras. Politicians championed them as a solution to the crisis of trust between law enforcement and communities of color. The technology was supposed to create an objective record that would protect both civilians and officers. It was the Panopticon made portable, strapped to every officer's chest.

The complications became apparent almost immediately. In most departments, officers retained the ability to activate and deactivate their cameras at will. Many departments had policies that required officers to begin recording before an encounter, but compliance was uneven and enforcement was rare. Officers involved in controversial incidents frequently claimed that their cameras had malfunctioned or that they had forgotten to turn them on. The footage was stored by the department itself, which created an obvious conflict of interest. When footage supported the officer's account, it was released quickly. When footage contradicted the official narrative, it was withheld for months under the claim that releasing it would compromise an ongoing investigation.

A major randomized controlled trial in Washington, D.C., published in 2019 by researchers David Yokum, Anita Ravishankar, and Alexander Coppock, complicated the picture further. The study was far larger and more rigorous than the Rialto experiment. It found no statistically significant effect of body cameras on police use of force or on civilian complaints.

- The cameras did not make officers measurably more restrained.
- The cameras did not make civilians measurably more cooperative.

The study suggested that the Rialto results might have been influenced by the Hawthorne effect, where subjects change their behavior simply because they know they are part of a study, rather than because of the cameras themselves.[4]

The lesson was clarifying. The camera only functions as a check on power when the person wearing it cannot control it. An officer who can turn the camera off at will is not truly being watched. He is performing the appearance of being watched while retaining the ability to become invisible at the crucial moment. The ring of Gyges does not require literal invisibility. It only requires the ability to choose when one is seen and when one is not.

Meanwhile, far from American police departments, citizen journalism was demonstrating the power of sousveillance on a global scale. During the Arab Spring of 2011, protesters across the Middle East and North Africa used their phones to document security forces firing on unarmed demonstrators. The footage was uploaded to YouTube and shared through social media platforms. It reached international media outlets that had been barred from entering the countries where the protests were occurring. The cameras in the hands of ordinary citizens became the primary source of information about revolutions that toppled governments.[9]

In Hong Kong between 2019 and 2020, protesters refined the practice into a sophisticated real-time documentation system. Live-streamers broadcast police actions as they happened. Footage was archived, cataloged, and distributed through encrypted channels. The protesters understood that their cameras were both weapons and shields. The footage deterred some police violence by making it immediately visible to a global audience. But the cameras also became tools of identification. Authorities used the same footage to identify protesters, track their movements, and build cases for prosecution. The watchers discovered that when everyone is watching everyone, the power of the camera flows in every direction at once.

On a May morning in 2020, a woman named Amy Cooper was walking her dog off-leash in a wooded section of Central Park called the Ramble. A Black man named Christian Cooper, no relation, asked her to leash the dog as the park's rules required. Their exchange escalated. Amy Cooper called 911 and told the dispatcher, with apparent distress in her voice, that "an African American man" was "threatening my life." Christian Cooper had recorded the encounter on his phone. He posted the video online.

The video went viral within hours. By the following day, Amy Cooper had been identified by name. She was fired from her job at an investment firm. She was charged with filing a false police report. She became the subject of thousands of articles, social media posts, and television segments. The internet had rendered its verdict before any investigation had begun.

The Amy Cooper incident was not an isolated event. It was part of a phenomenon that had been accelerating since the widespread adoption of smartphones and social media. The pattern repeated itself with mechanical regularity. Someone would behave badly in a public space. A bystander would record it. The video would be uploaded to social media. Within hours, internet users would identify the person in the video. Within a day, the person would lose their job, receive death threats, and become a permanent fixture in the digital record of public shame. The term "Karen" entered the cultural lexicon as shorthand for this specific type of recording. A white person, usually a woman, caught on camera behaving in a racially charged or entitled manner.

The recordings served a real social function. They exposed behavior that had previously been invisible. They demonstrated patterns of racial bias that many people denied existed. They gave marginalized communities a tool for documenting the casual cruelty they endured daily. The videos were evidence in the most literal sense, and the public response to them created a powerful deterrent against a certain kind of public misconduct.

But the mechanism had no calibration. The punishment was always the same regardless of the severity of the offense. A moment of poor judgment received the same treatment as an act of genuine malice. A person who said something thoughtless was subjected to the same global exposure as a person who committed a deliberate injustice. The journalist Jon Ronson documented this phenomenon in his book *So You've Been Publicly Shamed*, tracing the human cost of viral outrage through the stories of people whose lives were permanently destroyed by a single recorded moment.[6]

Ronson profiled Justine Sacco, a communications executive who posted an offensive joke on Twitter before boarding a flight to South Africa in 2013. By the time she landed, the tweet had gone viral. She had been fired. She had become the number one trending topic worldwide. She spent years trying to recover from the destruction of her professional and personal life over a single ill-considered sentence. The punishment bore no relationship to the offense. The mechanism of online shaming had no proportionality, no appeals process, and no expiration date.

The deeper problem was what scholars call context collapse. A video captures a sliver of reality. It records seconds or minutes of an encounter that may have lasted much longer. Viewers see what the camera shows them, but they do not see what happened before the recording started. They do not hear what was said off-camera. They do not know the history between the people involved. They do not understand the circumstances that led to the moment being filmed.

The act of recording itself shapes the narrative. The person who holds the camera decides when to start and when to stop. That editorial power is invisible to the viewer. A recording that begins mid-confrontation strips away the context that might explain how the confrontation started. A recording that ends before the resolution strips away the context that might show how it was resolved. The camera appears to provide objective truth, but the truth it provides is always partial. It is always framed by decisions the viewer cannot see.

The same event filmed from two different angles can tell two entirely different stories. One camera catches the moment a man shoves someone. A second camera, positioned differently, captures the shove that provoked it. A viewer who sees only the first recording concludes that the man is violent. A viewer who sees only the second recording concludes that the man was defending himself. Both viewers believe they have seen the truth. Neither has seen the whole picture.

This dynamic created an atmosphere of pervasive anxiety that reached far beyond the people captured in viral videos. Teachers reported growing reluctance to discipline students because they feared being recorded out of context and having the footage uploaded to social media. A teacher who raises her voice at a disruptive student can be made to look abusive when the recording omits the twenty minutes of escalating disruption that preceded the moment. Doctors described reluctance to have candid conversations with patients because recordings of those conversations could be extracted from context and used in lawsuits. Ordinary people reported hesitation to intervene in public conflicts because they feared being recorded, mischaracterized, and subjected to the same viral shaming they had seen destroy others.

The irony was precise and devastating. Universal watching was producing the same behavioral effects as the Panopticon that Jeremy Bentham had designed in the eighteenth century. When everyone is a potential recorder, everyone becomes a potential prisoner of their own worst moment. People do not become more virtuous under this pressure. They become more guarded. They withdraw from public engagement. They censor themselves not because they have developed better values but because they fear consequences that arrive without warning and without proportion. The camera was supposed to make people accountable. In practice, it was making people afraid.

In January 2018, the users of an internet forum called Reddit noticed something unusual. A series of pornographic videos had appeared online featuring the faces of well-known actresses grafted onto the bodies of performers. The videos were not crude cut-and-paste jobs. They were eerily convincing. The faces moved naturally. The expressions were fluid. The lip movements matched the audio. The creator used an artificial intelligence technique called a generative adversarial network to produce them. He posted under the username "deepfakes." The name stuck.

Within months, the technology proliferated. Open-source software made deepfake creation accessible to anyone with a computer and basic technical skills. The earliest versions required significant processing power and expertise. Each successive generation of the technology became cheaper, faster, and more convincing. By 2023, smartphone applications could generate a passable deepfake video in minutes. By 2025, the results were becoming indistinguishable from authentic footage to the untrained eye.

The implications for the camera phone's role as a check on power were catastrophic. Legal scholars Robert Chesney and Danielle Citron had identified the threat in a 2019 paper that introduced a concept they called the "liar's dividend." The term described a paradox that deepfakes created for the credibility of all video evidence. In a world where any video can be fabricated, the existence of deepfake technology gives every person captured on video a new and potent defense... the claim that the footage is fake.[7]

The liar's dividend does not require the accused person to produce a competing deepfake. It requires only the assertion that deepfakes exist, that the technology is widely available, and that the footage in question could therefore be fabricated. The burden of proof shifts from the accused to the accuser. Instead of asking "did this person do what the video shows?" the question becomes "can we prove this video is authentic?" The mere existence of the technology poisons the well of visual evidence.

Politicians were among the first to exploit the dividend. When authentic footage surfaced showing public officials making controversial statements or engaging in questionable conduct, the response increasingly followed a template. The official would declare the video a deepfake. Supporters would amplify the claim. The ensuing debate would shift from the content of the video to the authenticity of the video. By the time forensic analysis confirmed the footage was real, the news cycle had moved on. The denial had served its purpose. It had created enough doubt to neutralize the evidence.

The liar's dividend began infiltrating legal proceedings as well. Defense attorneys introduced deepfake claims to challenge video evidence in criminal cases. Even when forensic experts testified that the footage was authentic, the seed of doubt had been planted. Jurors who had seen convincing deepfakes in the news or on social media carried that awareness into the courtroom. The question "could this be fake?" became a reasonable consideration in a way it never had been before.

The epistemic crisis that deepfakes created struck at the foundation of everything the camera phone had achieved. The entire power of citizen recording rested on a single assumption that a video shows what happened. The Rodney King tape was powerful because viewers believed what they saw. Darnella Frazier's recording convicted Derek Chauvin because the jury believed the footage was an accurate representation of reality. Feidin Santana's video sent Michael Slager to prison because it proved that the officer had lied. All of that power depended on the credibility of the medium itself.

Deepfakes threatened to destroy that credibility entirely. If video could be fabricated convincingly enough that ordinary viewers could not distinguish real from fake, then video proved nothing. If video proved nothing, then the world reverted to the condition that had existed before George Holliday picked up his camcorder. It reverted to testimony. It reverted to one person's word against another's. And when a dispute comes down to testimony, institutional power wins. The officer's word outweighs the civilian's. The politician's denial outweighs the journalist's accusation. The powerful person's narrative prevails because power has always been the ultimate arbiter of whose story gets believed.

The ring was reasserting itself through technology. Deepfakes did not make people invisible. They did something more insidious. They made truth invisible. They created a world in which authentic evidence could be dismissed as fabrication and fabricated evidence could be presented as truth. The distinction between what happened and what people claimed happened was collapsing. The camera had been the most powerful tool ever devised for holding the powerful accountable. Deepfakes were turning it into a tool for manufacturing doubt.

An arms race emerged between detection and creation. Researchers developed forensic tools that could identify deepfakes by analyzing pixel patterns, lighting inconsistencies, and artifacts in facial movement. Each new detection method worked for a time. Then the generative models adapted. The fakes became better. They eliminated the specific artifacts that the detection tools were trained to find. New detection tools were developed. The cycle repeated itself.

Proposals for verification systems multiplied. Some researchers advocated for blockchain-based timestamping that would cryptographically certify when and where a video was recorded. Others proposed digital watermarking that would embed authentication data directly into footage at the moment of capture. Industry groups developed content provenance standards that would track the origin of media files through a chain of custody. Each proposal addressed a real problem. None had achieved widespread adoption. The fundamental difficulty was structural. It is always easier to create doubt than to establish certainty. A deepfake needs to be only convincing enough to trigger hesitation. A verification system needs to be universal, tamper-proof, and comprehensible to a jury. The asymmetry favored the forger.

The story of sousveillance is the story of a promise that was half kept. The promise was simple and powerful. If ordinary people could watch the powerful, the powerful would be forced to behave as though the rules applied to them.

But the promise was also half broken. The same technology that enabled accountability also enabled mob justice. Videos stripped of context destroyed innocent people. The viral shaming machine operated without proportionality, without due process, without mercy. The chilling effect of universal recording made people more guarded, not more honest. And then deepfakes arrived to threaten the entire edifice.

We have spent decades building tools to watch one another. We have not spent nearly enough time asking what we are supposed to do with what we see.

This is the paradox at the heart of the watching age. We can see everything now. We can record it, upload it, and share it with the world in seconds. The powerful can be exposed. The hidden can be revealed. The lies can be documented. But seeing is not the same as understanding. Recording is not the same as justice. The camera is the most powerful tool of accountability ever placed in ordinary hands, and it is not enough. It was never going to be enough. The problem is not that we cannot see what people do when they think no one is watching. The problem is what we become when we realize that everyone is.

References

[1] Cannon, Lou. *Official Negligence: How Rodney King and the Riots Changed Los Angeles*. Westview Press, 1999. See also *Koon v. United States*, 518 U.S. 81 (1996), and trial transcripts from *People v. Powell et al.*, No. BA035498 (Cal. Super. Ct. 1992).

[2] Smartphone ownership statistics, Pew Research Center, multiple years.

[3] Ariel, Barak, et al. "Contagious Accountability: A Global Multisite Randomized Controlled Trial on the Effect of Police Body-Worn Cameras on Citizens' Complaints Against the Police." *Criminal Justice and Behavior* 44, no. 2 (2017): 293-316.

[4] Yokum, David, Anita Ravishankar, and Alexander Coppock. "A Randomized Control Trial Evaluating the Effects of Police Body-Worn Cameras." *PNAS* 116, no. 21 (2019): 10329-10332.

[5] *State v. Chauvin*, Hennepin County District Court, 2021. Trial testimony of Darnella Frazier.

[6] Ronson, Jon. *So You've Been Publicly Shamed*. Riverhead Books, 2015.

[7] Chesney, Robert, and Danielle Keats Citron. "Deep Fakes: A Looming Challenge for Privacy, Democracy, and National Security." *California Law Review* 107, no. 6 (2019): 1753-1820.

[8] Mann, Steve, Jason Nolan, and Barry Wellman. "Sousveillance: Inventing and Using Wearable Computing Devices for Data Collection in Surveillance Environments." *Surveillance & Society* 1, no. 3 (2003): 331-355.

[9] Howard, Philip N., and Muzammil M. Hussain. *Democracy's Fourth Wave? Digital Media and the Arab Spring*. Oxford University Press, 2013.

Chapter Fourteen - The Lost Wallet

In the summer of 2015, a young research assistant walked into the lobby of a five-star hotel in Zurich. He carried a transparent wallet containing a grocery list, a key, three business cards bearing the name of a fictitious person, and the equivalent of thirteen U.S. dollars in local currency. He approached the front desk, set the wallet down, and told the receptionist he had found it on the street near the entrance. He asked if the receptionist could take care of it. Then he left.

The receptionist now faced a choice that no one would ever audit. The wallet's owner was not real. No camera in the lobby was tracking this particular item. No supervisor was watching the interaction. The research assistant was gone. The receptionist could pocket the money and toss the wallet in the trash, or could attempt to contact the owner listed on the business cards. There was no mechanism of accountability and no consequence either way.

This was the Ring of Gyges in miniature. Not a golden band pulled from a chasm in the earth, but a transparent sleeve of plastic dropped on a hotel counter. The question it posed was the same question Glaucon asked centuries ago in Plato's Republic. If you can take something with no chance of being caught, will you take it?

This chapter asks a different question. Not what anecdotes suggest, but what the data actually shows when researchers design controlled experiments to test Glaucon's hypothesis directly. The experiments are rigorous. The sample sizes are enormous. The results span continents, cultures, and economic conditions. The answer turns out to be more complicated than the cynics predicted. It also turns out to be more hopeful.

The Zurich hotel was one stop among thousands. The wallet drop was part of the largest field experiment on civic honesty ever conducted. Alain Cohn, Michel Andre Marechal, David Tannenbaum, and Christian Lukas Zund designed the study over the course of several years and published their results in the journal *Science* in 2019.[1] Their research assistants performed the same routine 17,303 times. They walked into banks, hotels, museums, post offices, police stations, and cultural institutions across 355 cities in 40 countries. Each time, the assistant approached an employee, presented the wallet, claimed to have found it nearby, and left.

The wallets came in three varieties. Some contained no money at all. Some contained the equivalent of roughly thirteen dollars in local currency. A smaller subset contained roughly ninety-four dollars. Every wallet included the same business cards, the same grocery list, the same key. The only variable was the cash. The design was elegant in its simplicity. By holding everything else constant, the researchers isolated a single question... does the presence of money change whether a stranger does the right thing?

Before the experiment began, the research team ran a parallel survey. They asked a group of professional economists to predict the results. They asked business professors. They asked over two thousand ordinary people from across the globe. The question was simple. As the amount of money in the wallet increases, will people be more likely or less likely to return it?

The prediction was nearly unanimous. Everyone expected the same pattern. More money means more temptation. More temptation means fewer returns. The logic was straightforward and it tracked perfectly with rational self-interest theory, the same framework that has dominated economic thinking for more than a century. A wallet with ninety-four dollars offers a bigger reward for dishonesty than a wallet with thirteen. A wallet with no money offers no reward at all. Therefore, empty wallets should be returned most often and full wallets least often. The prediction was intuitive. It felt like common sense. It aligned with every model of human behavior that treats self-interest as the default setting of the species.

The economists were confident. The business professors were confident. The general public was confident. Across every demographic surveyed, the consensus held without exception. Homo economicus would behave rationally, and rational behavior meant pocketing the cash.

They were all wrong.

In 38 of 40 countries, wallets containing money were returned at higher rates than empty wallets.[1] The finding was not marginal. It was decisive. Across nearly every country, every type of institution, and every cultural context, the presence of cash made people more honest rather than less. The researchers then introduced the high-value wallets containing ninety-four dollars. The return rate climbed even higher.[1]

The pattern held in wealthy nations and poor ones. It held in countries ranked high on corruption indices and countries ranked low. It held in Scandinavia, where trust between strangers is among the highest in the world. It held in parts of Africa, Southeast Asia, and Latin America. Only two countries showed a slight reversal of the trend, and even those results were not statistically robust. The overall picture was remarkably consistent. Give people more reason to steal, and they become less likely to do so.

The research team proposed an explanation rooted in the psychology of self-image. When a receptionist finds an empty wallet, keeping it feels like negligence rather than crime. There is nothing in it. Throwing it away or ignoring it carries no moral weight. But when money is present, the calculus shifts. Keeping the wallet with cash inside feels like theft. The receptionist is no longer just failing to act. The receptionist is actively taking something that belongs to someone else. The psychological cost of seeing yourself as a thief outweighs the financial benefit of thirteen dollars or even ninety-four.

Cohn and his colleagues called this "theft aversion."[1] The term describes a reluctance not to lose money but to acquire the identity of someone who steals. Most people, it turns out, will forfeit a modest financial gain to avoid the internal experience of being dishonest. The reward is real, but the self-concept is worth more.

This was the Ring of Gyges tested at global scale, and the results contradicted Glaucon's prediction. The ring was available. The hotel receptionists and bank clerks and museum workers could have pocketed the cash without consequence. Most of them chose not to.

The finding demands a caveat. The stakes in the wallet study were low. Thirteen dollars is a cup of coffee and a sandwich. Ninety-four dollars is a nice dinner. Nobody was offered a year's salary in a transparent sleeve. The ring's power may scale with the prize, and nothing in the wallet study tells us what would happen if the amount were ten thousand dollars or a hundred thousand. Glaucon's shepherd did not find a few coins in a cave. He found the power to become king.

Still, the wallet study revealed something that economic models had not predicted and that philosophical cynicism had not allowed for. At the level of ordinary daily life, most people possess a moral instinct that activates precisely when the opportunity for dishonesty presents itself. The temptation does not override the conscience. In most cases, it awakens the conscience. The ring was sitting on the counter in 355 cities across 40 countries. Thousands of people looked at it and walked away.

Twenty years before the wallet study, a man named Paul Feldman arrived at a similar conclusion through a very different method. Feldman was an economist who worked at the American Institute for Research in Washington, D.C. In the early 1980s, he started bringing bagels to work. He would carry a bag of fresh bagels into the office break room, set them on a table, and leave a small basket beside them with a hand-written sign, take a bagel, leave a dollar.

His colleagues took bagels and left dollars. Not always, but often enough to cover his costs. Feldman found the ritual satisfying in a way that his desk work was not, and in 1984 he quit his job to become a bagel deliveryman. He began servicing office buildings throughout the Washington metropolitan area, delivering bagels and collection baskets each morning to dozens of companies. Each day he recorded exactly how many bagels he left and exactly how much money he collected.[2]

He did this for eighteen years. Every weekday morning, he loaded his car with fresh bagels, drove his route through the office parks and government buildings of the capital region, and left breakfast on honor for thousands of strangers. The business model was absurdly simple and utterly dependent on human decency.

The data set Paul Feldman accumulated was extraordinary in its breadth and granularity. Over nearly two decades, he delivered bagels to roughly 140 companies each week. He recorded the weather, the day of the week, the size of the office, the time of year, and the payment rate for each delivery. He was not an academic researcher running a funded study. He was a bagel man running a business on the honor system. But the data he generated was as clean as anything produced in a university laboratory.

The overall payment rate across all companies and all years was approximately 87 percent.[2] For every hundred bagels Feldman left on counters in office break rooms, about eighty-seven people paid the full dollar. Most people, confronted with a small and perfectly anonymous opportunity to eat without paying, chose to pay.

The remaining thirteen percent were not uniform. Feldman's records revealed patterns that illuminated the architecture of everyday dishonesty. Weather affected honesty. On pleasant days, payment rates rose. On cold, rainy, or snowy days, they fell. The connection was not dramatic, but it was consistent enough to track across thousands of data points.[2] People in bad moods were slightly more likely to take something for nothing.

Office size mattered more than weather. Small offices with fewer than two dozen employees showed significantly higher payment rates than large offices with hundreds of workers.[2] In a small office, you might be the only person in the break room. If a bagel disappears without payment, your colleagues might notice. In a large office, you are one face among many. The anonymity of the crowd offers a kind of cover, and the payment rate dropped accordingly. The larger the group, the more invisible each individual becomes, and the more license that invisibility grants.

Feldman also discovered a correlation that delighted the journalists Steven Levitt and Stephen Dubner when they documented his story in their 2005 book *Freakonomics*. Executive floors paid at lower rates than floors populated by lower-level employees.[2] The people at the top of the corporate hierarchy, the ones with the highest salaries and the most authority, were more likely to take a free bagel. The ones with less money and less power were more likely to pay.

This finding resonates with a pattern that has surfaced throughout this book. Power and anonymity interact in predictable ways. The executives were not more anonymous than the support staff. They were more accustomed to receiving benefits without personal cost. Their position in the hierarchy had normalized a particular relationship to entitlement. The bagel was simply one more thing the organization owed them.

Holidays created their own distortions. Payment rates dropped before Thanksgiving and Christmas, rose slightly after New Year's Day, and dropped again during tax season.[2] Stressful periods at individual companies produced measurable dips. When a firm announced layoffs, the bagel basket came up short for weeks. The data painted a portrait of moral behavior as something almost meteorological, subject to fronts, to pressure systems, to invisible forces that shift the atmosphere of a workplace without anyone noticing the change.

The beauty of Feldman's data was its ordinariness. He was not studying dramatic moral choices. He was studying the daily micro-decisions of office workers who wanted breakfast. The bagel was worth a dollar. The question of whether to pay for it was trivial by any measure. Yet the patterns that emerged from eighteen years of trivial questions illuminated something profound about the conditions under which people choose honesty over opportunism.

Paul Feldman's data told a story that was fundamentally different from the one Glaucon imagined. It was also different from the one Socrates defended. Feldman's customers were neither purely selfish nor purely virtuous. They were something in between. They were people whose honesty fluctuated with context, mood, group size, and seasonal stress. Morality was not a fixed trait in these office workers. It was a variable, responsive to environmental conditions the way a thermostat responds to temperature.

Glaucon argued that people are only honest because they fear punishment. Feldman's data suggested otherwise. There was no punishment for not paying. Nobody was watching. Nobody was checking. Eighty-seven percent paid anyway. But Socrates argued that the just person is just in all circumstances. Feldman's data suggested otherwise on that front as well. Thirteen percent of the time, the bagel vanished without payment. On bad days in large offices, the rate climbed higher. The conscience was real, but it was not invincible.

Laboratory experiments have probed this middle ground with increasing precision over the past three decades. The most famous is the dictator game, a deceptively simple exercise that has been replicated thousands of times across dozens of countries. The setup is straightforward. A researcher gives one participant a sum of money, typically ten dollars. This participant is designated the "dictator." The dictator can share any portion of the money with a second participant, from zero to the full amount. The second participant must accept whatever is offered. There is no negotiation and no retaliation. The dictator holds all the power.

Rational self-interest theory makes a clear prediction about this game. The dictator should keep everything. There is no strategic reason to share. The second player cannot punish the dictator for greediness. The optimal move, by the logic of pure self-interest, is to offer nothing.

Christoph Engel conducted a meta-analysis of more than 130 dictator game studies in 2011.[4] The results showed that the average dictator offers between 20 and 30 percent of the total. Many participants split the money evenly. Very few keep everything. Even when anonymity is maximized, so that the dictator believes no one will ever learn of the decision, offers decrease but do not collapse to zero.[4] Some participants share generously even when they are completely invisible.

The dictator game reveals the presence of an internal fairness instinct that operates independently of social pressure. People give money away when they have no reason to, no audience to impress, and no punishment to fear. The instinct is not universal. Some dictators do keep everything, and the researchers who study these games have learned to expect a range of behavior that spans from full generosity to full exploitation. But the consistent finding across hundreds of studies is that pure selfishness is the minority behavior, not the norm. The average person, left alone with total power and total anonymity, still shares. Not equally. Not generously by most standards. But the offer is not zero, and that non-zero number represents something that rational self-interest theory cannot explain.

Dan Ariely, a behavioral economist at Duke University, pushed the investigation further with a series of experiments documented in his 2012 book *The (Honest) Truth About Dishonesty*.[3] Ariely's signature method was the matrix task. Participants received a sheet of paper containing twenty simple math problems. They had five minutes to solve as many as possible, and they earned a small cash reward for each correct answer.

The twist was simple and devastating. In the experimental condition, participants shredded their worksheets before reporting their scores. They walked to a paper shredder in the corner of the room, fed their answer sheets into it, and then told the researcher how many problems they had solved. Nobody could check. The evidence was gone. The opportunity to cheat was absolute, and the participants knew it.[3]

Most people cheated. That finding alone might seem to validate Glaucon. But the pattern of cheating told a different story. The average participant inflated their score by roughly two problems. They did not claim to have solved all twenty. They did not even claim to have solved fifteen. They nudged their real score up by a small amount, just enough to earn a few extra dollars.[3]

Ariely ran variations on the experiment across multiple contexts. He changed the reward amounts. He changed the difficulty of the problems. He changed the conditions under which participants reported their scores. He tested participants in different countries and different demographic groups. The core finding remained stable across every variation. People cheated a little. Almost nobody cheated a lot. The maximum possible gain sat there, fully available, completely unmonitored. Participants left most of it on the table.

Ariely proposed that humans maintain what he called a "personal fudge factor."[3] Most people want to benefit from dishonesty. They also want to see themselves as honest. These two desires create a tension, and the resolution is a compromise. People cheat enough to gain a small advantage, but not so much that they are forced to update their self-image. They bend the truth rather than break it. They pick up the ring and try it on, but they do not use it to walk into the treasury.

The fudge factor explains something that neither Glaucon nor Socrates fully anticipated. Glaucon predicted that people wearing the ring would exploit it to the maximum. He imagined a shepherd who used invisibility to seduce a queen and murder a king. Socrates predicted that truly just people would not exploit the ring at all. He imagined a soul so well-ordered that temptation simply could not reach it. The experimental evidence suggests that most people occupy a middle position that neither philosopher described. They exploit the ring partially, extracting modest benefits while preserving the internal narrative that they are fundamentally good. The ring at partial power, used cautiously, used with one eye on the mirror.

A related line of research has explored how this internal narrative interacts with prior behavior. In 2010, Anna Merritt, Daniel Effron, and Benoit Monin published a review of what psychologists call moral licensing.[8] The phenomenon is straightforward. When people have recently done something virtuous, they become more likely to act selfishly in subsequent decisions. A person who donates to charity in the morning may tip less generously at dinner. An employee who helps a colleague with a difficult task may feel entitled to cut a corner the following day.

The mechanism is a kind of moral accounting. Good deeds create credits in a psychological ledger. Bad deeds create debits. Most people aim for a balance that stays in the black, but they do not aim for maximum virtue.[8] They aim for a comfortable surplus, enough goodness to maintain self-respect without demanding sainthood.

Moral licensing suggests that the human moral system is not a compass pointing consistently toward true north. It is a thermostat, regulating behavior around a set point. When the temperature of recent good behavior rises, the system permits a slight cooling. When the temperature drops because of a recent transgression, the system pushes toward correction. The result is a kind of moral homeostasis, a fluctuation around a mean rather than a steady trajectory in any one direction.

The experiments described so far measure behavior. They tell us what people do when faced with anonymous moral choices. They do not tell us what is happening inside the brain when those choices are made. A separate body of research has investigated the neuroscience of moral decision-making, and the findings add a critical layer to our understanding of the ring's effect.

Joshua Greene, a philosopher and neuroscientist at Harvard University, conducted a series of landmark studies in the early 2000s using functional magnetic resonance imaging to observe the brain in real time as participants wrestled with moral dilemmas.[5] The dilemmas Greene used were variations on the famous trolley problem. In one version, a participant imagines standing by a switch. A runaway trolley is headed toward five people. Pulling the switch diverts the trolley to a side track where it will kill one person. Most people say pulling the switch is morally acceptable.

In another version, the participant imagines standing on a bridge above the tracks. The only way to stop the trolley is to push a large stranger off the bridge and into its path. The outcome is identical. One person dies instead of five. But most people say pushing the stranger is morally unacceptable.[5]

Greene's brain scans revealed the reason for this inconsistency. When participants considered the impersonal dilemma of pulling the switch, the dorsolateral prefrontal cortex activated strongly.[5] This brain region is associated with rational calculation, abstract reasoning, and cost-benefit analysis. The brain treated the switch scenario as a math problem. Five lives outweigh one.

When participants considered the personal dilemma of pushing the stranger, a different region activated with far greater intensity. The ventromedial prefrontal cortex, along with the amygdala and other areas associated with emotional processing, lit up on the scans.[5],[6] The brain was no longer calculating. It was recoiling. The prospect of physically harming another person triggered an emotional alarm that no amount of utilitarian arithmetic could silence in most subjects.

Greene's conclusion was that human moral cognition involves two distinct systems operating simultaneously.[6] One system is rational and calculative. It weighs outcomes and maximizes utility. The other is emotional and instinctive. It responds to the felt proximity of harm. The two systems often agree, but when they conflict, the result is the kind of moral uncertainty that most people experience as the sensation of being torn. The discomfort of moral dilemmas is not confusion. It is the sound of two neural systems arguing with each other, each equipped with its own logic, each capable of overriding the other under different conditions.

This finding matters for the Ring of Gyges because it establishes that morality is not a single faculty that either functions or fails. It is a negotiation between competing neural processes, each with its own evolutionary history and its own triggers. When Glaucon argued that people are only moral because of social consequences, he was describing only the rational system. Remove the consequences, and the rational system has no reason to restrain self-interest. But the emotional system does not depend on consequences. It responds to the act itself. It recoils from harm not because harm is punished but because the brain is wired to register harm as aversive. This wiring may be the product of millions of years of social evolution, built into primate brains that survived by cooperating as much as by competing.

A second line of neuroscience research addressed the specific question of what changes in the brain when a person believes they are being observed. This is the question at the heart of the Ring of Gyges, translated into the language of neuroscience. Keise Izuma, along with Daisuke Saito and Norihiro Sadato, published research in 2008 examining how the brain processes social reward under conditions of visibility versus invisibility.[7]

Izuma's team found that when participants believed someone was watching them make a generous choice, the striatum activated.[7] The striatum is the brain's primary reward center. Being seen while doing good produced a measurable neurological reward, a burst of the same activation associated with receiving money, eating delicious food, or hearing a compliment. Generosity in public felt good in a concrete and biological sense.

When participants believed they were unobserved, the pattern shifted. Generosity still occurred, but the striatum's contribution diminished. Instead, the prefrontal cortex carried a greater share of the processing load.[7] The brain was working harder. Making the moral choice without an audience required more cognitive effort because the easy reward of social approval was absent. The person had to generate the motivation internally rather than receiving it from the environment. This difference was visible on the scans. The brain of a person choosing generosity in private looked measurably different from the brain of a person choosing generosity in public. Both arrived at the same behavior through different neural pathways.

Two moral systems emerged from this research, and they mapped cleanly onto the debate between Glaucon and Socrates. The first system is reputation-based. It operates when we are watched. It makes virtue feel effortless because the social environment rewards it. The second system is internally generated. It operates when we are alone. It makes virtue feel like work because the brain must supply its own motivation without external reinforcement.

We can now assemble the evidence from wallets, bagels, dictator games, matrix tasks, moral licensing studies, and brain scans into a composite answer to Glaucon's question. The evidence comes from forty countries. It comes from thousands of office break rooms and hundreds of university laboratories. It comes from fMRI machines that photograph the brain in the act of moral reasoning. The answer is not clean. It does not fit on a bumper sticker. It is precisely because it resists simplification that it should be taken seriously.

At one end of the spectrum sits a small minority who will always exploit the ring fully. These are the true Gyges figures, the people who take every dollar from the dictator game, who claim to have solved every problem on the matrix task, who would keep the wallet and the cash and never think about it again. They exist. The data confirms their existence. They are not the majority.

At the other end sits a small minority who will never exploit the ring at all. These are the Socratic figures, the people who split the money evenly even under total anonymity, who report their scores honestly even when cheating is guaranteed to be undetectable, who would track down the wallet's owner and hand-deliver it. They exist too. The data confirms their existence as well. They are also not the majority.

The majority sits in between. They are influenced by context, by incentives, by community, and by habit. They will pay for the bagel on most days but not all. They will return the wallet when it contains enough money to activate their theft aversion but might not bother when it contains nothing. They will share some money in the dictator game but keep most of it. They will cheat a little on the matrix task but not a lot.

These people are not moral failures. They are moral humans. Their virtue is real but conditional. Their selfishness is real but bounded. They navigate the space between Glaucon's cynicism and Socrates' idealism, making decisions that are shaped by the world around them as much as by the principles within them. They are, in other words, most of us. The great middle of the human moral distribution is not a wasteland between two peaks. It is the terrain where character actually lives, responsive to gravity, to weather, to the landscape through which it moves.

This is the most important finding in this book, and it has been hiding in plain sight throughout every chapter. The answer to Glaucon's question is neither "everyone would use the ring" nor "no one would." The answer is that it depends. It depends on the size of the temptation. It depends on the mood of the person facing it. It depends on whether the person is in a small office or a large one, whether the day is sunny or overcast, whether the holidays are approaching or receding. It depends on whether the person recently did something good or something selfish. It depends on the culture they grew up in, the habits they have formed, and the communities that shaped them.

What matters about this answer is not its complexity. What matters is what follows from it. If moral behavior were purely a product of surveillance, as Glaucon believed, then the only way to ensure virtue would be total observation. We explored that possibility in Chapter 12. The result was compliance, not goodness. If moral behavior were purely a product of the soul, as Socrates believed, then no external intervention would be necessary. We have numerous chapters of evidence that this is not the case.

But if moral behavior is influenced by context, then context is a lever. If honesty rises in smaller communities, then community design matters. If generosity increases when people are reminded of moral norms, then those reminders matter. If cheating decreases when people feel connected to a shared identity, then the strength of that shared identity matters. The variability of virtue is not a weakness in the human moral system. It is a feature, because it means the system is responsive. It means that virtue can be cultivated, reinforced, and supported by the structures we build around ourselves. The question is no longer whether people are inherently good or inherently selfish. The question is what conditions produce which outcome, and whether we have the wisdom to create the right ones.

The wallet study's most striking finding wasn't that most people returned wallets. It was that the presence of money activated rather than suppressed the moral instinct. The temptation itself became the trigger for honesty. This suggests that the Ring of Gyges does not merely test character. It also reveals character. It shows us something about ourselves that ordinary, observed daily life keeps hidden. When no one is watching, most people discover that they do not want to be thieves. The discovery is not effortless. It requires the brain to work harder than it does under observation. But the work gets done.

The next two chapters explore what that alteration looks like in practice. If the ring's power depends on context, then the question shifts from "Are people good?" to "What kind of world makes people better?" Glaucon framed the question as a test of individual character. The evidence reframes it as a question about collective design.

Plato's shepherd found the ring alone in a cave. We carry ours in our pockets, surrounded by the societies we have built. What those societies look like determines what we do with the ring far more than what we are born with determines it. Seventeen thousand wallets were dropped across the surface of the earth. Most of them came back. The story of what that means, and what we can build on top of it, begins in the next chapter.

References

[1] Cohn, Alain, Michel André Maréchal, David Tannenbaum, and Christian Lukas Zünd. "Civic Honesty Around the Globe." *Science* 365, no. 6448 (2019): 70-73.

[2] Levitt, Steven D. and Stephen J. Dubner. *Freakonomics*. William Morrow, 2005.

[3] Bazerman, Max H., and Ann E. Tenbrunsel. *Blind Spots: Why We Fail to Do What's Right and What to Do About It*. Princeton University Press, 2011.

[4] Engel, Christoph. "Dictator Games: A Meta Study." *Experimental Economics* 14, no. 4 (2011): 583-610.

[5] Greene, Joshua D. et al. "An fMRI Investigation of Emotional Engagement in Moral Judgment." *Science* 293, no. 5537 (2001): 2105-2108.

[6] Greene, Joshua D. *Moral Tribes: Emotion, Reason, and the Gap Between Us and Them*. Penguin, 2013.

[7] Izuma, Keise, Daisuke N. Saito, and Norihiro Sadato. "Processing of Social and Monetary Rewards in the Human Striatum." *Neuron* 58, no. 2 (2008): 284-294.

[8] Merritt, Anna C., Daniel A. Effron, and Benoît Monin. "Moral Self-Licensing: When Being Good Frees Us to Be Bad." *Social and Personality Psychology Compass* 4, no. 5 (2010): 344-357.

Chapter Fifteen - Building People Who Don't Need the Ring Taken Away

There is a question that pollsters have been asking people in dozens of countries for more than four decades. The question is simple and the answer reveals more than almost any other single data point in the social sciences. It goes like this "Generally speaking, would you say that most people can be trusted, or that you need to be very careful in dealing with people?"

The World Values Survey has posed this question to hundreds of thousands of respondents across nearly one hundred nations since the early 1980s. [1] The results, mapped across continents and decades, produce something remarkable. They produce a portrait of the human species sorted by its own faith in itself.

In Denmark, roughly sixty-seven percent of people say that most people can be trusted. In Norway, the number hovers around sixty-five percent. In Sweden, it sits near sixty-three percent. Finland follows close behind.[1] These are countries where a majority of the population looks at a stranger and assumes good faith. The default setting for most Scandinavians is trust.

Now move south. In Brazil, about seven percent of the population says that most people can be trusted. In Colombia, the figure dips to four percent.[1] In these countries the default setting is suspicion. The assumption is that a stranger will take advantage if given the chance. The ring, in Glaucon's terms, is always on someone's finger.

The United States sits in the middle of this spectrum, and it is moving in the wrong direction. In 1972, roughly forty-five percent of Americans said most people could be trusted. By the early 2020s, that number had fallen to approximately thirty percent.[1] A country that once sat comfortably in the upper tier of trusting nations has spent half a century sliding toward the suspicious end of the scale. The decline has not been sudden or dramatic. It has moved like erosion, a percentage point here and a percentage point there, so gradual that no single event explains it and no single generation can remember when it started. Americans trust each other less than their parents did, and their parents trusted each other less than their grandparents did. The trend line bends in only one direction.

The OECD (Organization for Economic Co-operation and Development) tracks a related but distinct measure. It asks citizens whether they trust their own governments, their courts, their police forces, their health systems.[2] The results correlate with the interpersonal trust data in ways that are almost too neat. Countries where people trust each other also tend to trust their institutions. Countries where people fear their neighbors also tend to fear their governments. Trust, it turns out, is not divisible. It operates as a system.

The obvious question is what separates the high-trust societies from the low-trust ones. Researchers have spent decades trying to answer it, and the list of contributing factors is long. But four variables appear consistently at the top of nearly every analysis.

The first is income inequality. Richard Wilkinson and Kate Pickett demonstrated in their landmark study *The Spirit Level* that societies with smaller gaps between their richest and poorest citizens display higher levels of trust across virtually every measurable dimension.[3] The Gini coefficient, which measures income distribution on a scale from zero to one, correlates with trust more reliably than almost any other single variable. When the gap between the top and the bottom is narrow, people trust each other more. When the gap widens, trust erodes.

The second factor is the social safety net. Countries that provide universal healthcare, robust unemployment insurance, publicly funded education, and reliable pension systems produce citizens who are less desperate and therefore less likely to view every interaction as a potential threat.[2] When your survival does not depend on winning a competition against your neighbor, you can afford to assume your neighbor is not trying to destroy you.

The third factor is institutional transparency. Countries where government officials are held accountable, where corruption is prosecuted, and where public spending is visible to citizens generate more trust than countries where power operates behind closed doors.[2] This is not a coincidence. When institutions behave honorably in the light, people learn to extend that assumption into the dark.

The fourth factor was once considered the most important, but recent data has complicated the picture. Ethnic and cultural homogeneity was historically correlated with high trust, and the Scandinavian countries that top the trust rankings are indeed relatively homogeneous societies.[1] But Canada ranks high on trust surveys despite being one of the most ethnically diverse nations on earth. New Zealand scores well despite significant Maori, Pacific Islander, and immigrant populations. The evidence now suggests that homogeneity matters less than equality. A diverse society where income gaps are small and institutions function fairly can generate high trust. A homogeneous society with vast inequality and corrupt institutions will not.

The mechanism behind all of this connects directly to the ancient thought experiment at the center of this book. When people's basic needs are met, when institutions function fairly, when the gap between rich and poor is manageable, the incentive to exploit the ring decreases. Gyges was a shepherd. He was a man with nothing, tending flocks that belonged to someone else, living at the bottom of a hierarchy he did not choose. The ring offered him everything he lacked. But what if he had already had enough? What if his society had provided him with security, dignity, and a genuine stake in the community? The ring would still have granted invisibility. It would still have offered power without consequence. But the desperation that drove a shepherd to murder a king and seize a throne might not have been there to exploit.

The high-trust societies of the world have not solved Glaucon's problem by constructing a better argument against it. They have solved it by building systems that reduce the pressure to exploit it. They have not made the ring less powerful. They have made the ring less necessary.

This distinction matters more than it might first appear. Surveillance is one response to the ring. It says... we know you would use the ring, so we will watch you to prevent it. Punishment is another response. It says... we know you would use the ring, so we will hurt you if you do. Both of these responses accept Glaucon's premise. They assume that human beings are fundamentally self-interested and must be controlled. The high-trust approach rejects the premise entirely. It says... if we build the right conditions, most people will not want to use the ring in the first place. The question is not how to stop people from being bad. The question is how to build a world where being good is the easier choice.

The road to Halden Prison winds through pine forests in southeastern Norway, about a hundred kilometers from Oslo. When you arrive, the first thing you notice is what is missing. There is no barbed wire. There are no watchtowers. The perimeter wall exists, but it is designed to blend into the landscape rather than dominate it. The buildings look like the campus of a well-funded Scandinavian architecture school, all clean lines and natural wood and floor-to-ceiling windows.[4]

Halden opened in 2010. It is a maximum-security facility, the highest security classification in the Norwegian system. The people inside have committed some of the most serious crimes in Norway, including murder, armed robbery, and sexual assault. These are not minimum-risk offenders selected for good behavior, nor white-collar criminals serving short sentences in comfortable conditions. They are the most dangerous prisoners in the Norwegian system, and the facility was designed specifically for them.

They live in private rooms. Each room has a flat-screen television, a small refrigerator, a desk, and a window that lets in natural light.[4] The rooms are not called cells. The doors have locks, but the aesthetic is closer to a college dormitory than to anything an American would recognize as a prison. Down the hall from the rooms are shared kitchens where inmates prepare their own meals with knives and utensils that would be considered contraband in virtually any American correctional facility.

The rest of the facility is equally disorienting for anyone raised on American assumptions about incarceration. The grounds include a recording studio where inmates produce music, a woodworking shop with industrial-quality equipment, a climbing wall, jogging paths through the trees, and a grocery store where inmates shop with a form of internal currency.[4]

The guards at Halden undergo two years of training before they set foot in the facility. That training covers psychology, ethics, conflict resolution, and human rights law.[4] In the United States, most correctional officers receive a few weeks of training, much of it focused on physical control techniques. The Norwegian model invests two years in preparing guards to be something more than enforcers. Once inside, they do not simply patrol corridors and count heads. They eat meals with inmates, play sports alongside them, and engage in conversation about goals, about families, about what went wrong and what might go right. The relationship is deliberately structured to resemble something closer to mentorship than to surveillance.

The philosophy behind all of this can be summarized in a single question that Are Hoidal, Halden's first warden, repeated in virtually every interview he gave. "Every inmate is going to be released. What kind of person do you want walking out that door?"[4]

That question contains an implicit theory of human nature, and it stands in direct opposition to the theory that drives American corrections. The American theory, broadly stated, is that criminals are people who made bad choices and need to be punished until they stop making them. The Norwegian theory is that criminals are people whose moral development was disrupted and who need to be placed in an environment where it can resume.

The difference between these two theories is not abstract. It produces measurably different outcomes in the lives of real people. Norway's recidivism rate hovers around twenty percent. Roughly one in five people released from Norwegian prisons reoffends within two years.[5] The United States Bureau of Justice Statistics reports that approximately seventy-six percent of released prisoners are rearrested within five years.[6] The measurement periods differ, two years versus five years, and the metrics differ, reconviction versus rearrest. Even after adjusting for these differences, the gap is enormous. Norway releases a person from prison and four out of five times that person stays out. America releases a person from prison and three out of four times that person comes back. Norway gets its citizens back. America cycles them through a revolving door that costs taxpayers billions of dollars a year while making no one safer.

The philosophical framework underpinning the Norwegian approach is explicitly Aristotelian, though the administrators would not always use that word. Aristotle argued in the *Nicomachean Ethics* that virtue is not a piece of knowledge you can simply hand to someone.[12] You cannot lecture a person into being just. Virtue is a habit, a practiced capacity that develops through repetition within a supportive community. A person becomes courageous by practicing courage in situations that require it, honest by practicing honesty when dishonesty is available, trustworthy by living in an environment that extends trust and then proving worthy of it.

Halden is designed around this principle. Inmates are given responsibility because responsibility builds the capacity for more responsibility. They are trusted with kitchen knives because learning to be trustworthy with a knife is part of learning to be trustworthy in general. They live in rooms with windows because a person who wakes up to natural light and makes their own breakfast is practicing the habits of a free citizen, and one day they will need those habits to function outside.

The counterargument is obvious and it deserves a direct answer. Norway is a small country, wealthy and homogeneous, with a population roughly equivalent to the state of Minnesota. It has universal healthcare, generous unemployment benefits, and a social safety net that catches people before they fall far enough to commit crimes of desperation. The United States has three hundred and thirty million people, staggering inequality, a minimal safety net, and a criminal justice system shaped by centuries of racial injustice. The conditions are not comparable.

This is true, and it would be dishonest to pretend otherwise. Halden cannot simply be replicated in Texas or Mississippi. The population dynamics are different. The economic base is different. The political culture is different. The history is different in ways that matter profoundly.

But the philosophy can travel even when the specific model cannot. Several American facilities have begun experimenting with Norwegian-inspired approaches. The Vera Institute of Justice has documented programs that incorporate restorative practices, reduced use of solitary confinement, and increased emphasis on education and vocational training.[5] The early results are promising, though the sample sizes remain small. The question is not whether America can become Norway. The question is whether the principle that treats inmates as people capable of moral growth produces better outcomes than the principle that treats them as problems requiring containment. The data from every country that has tried both approaches points in the same direction.

The ring of Gyges gives its wearer the power to act without consequence. Halden Prison represents an attempt to build people who would choose not to wear it, not because they fear punishment or because someone is watching, but because they have been given the opportunity to develop into the kind of person for whom the ring holds less appeal.

There is another way to address the harm the ring causes, and it does not involve preventing future crimes. It involves repairing the damage already done. The concept is called restorative justice, and its core premise can be stated as a shift in the question that drives a legal system. Traditional criminal justice asks... what law was broken and how should the offender be punished? Restorative justice asks... who was harmed and how can that harm be repaired?[7]

The distinction sounds subtle. In practice, it transforms everything. The roots of restorative justice extend back centuries into indigenous cultures around the world. The Maori of New Zealand practiced community-based resolution of harm long before Europeans arrived. First Nations communities across Canada maintained similar traditions. Aboriginal Australians developed elaborate systems of communal accountability and repair that predated Western legal concepts by millennia.[7] These were not primitive alternatives to "real" justice. They were sophisticated systems built on a fundamentally different understanding of what crime does to a community.

Western legal systems began to formalize restorative practices in the 1970s, partly through the influence of indigenous advocates and partly through a growing body of research suggesting that conventional prosecution was failing to achieve its stated goals. Victims emerged from courtrooms feeling ignored. They had been witnesses to their own suffering rather than participants in its resolution. Offenders served their sentences and returned to the same circumstances that produced the crime. Communities watched the cycle repeat. The system was efficient at processing cases and imposing sentences, but it was remarkably bad at the thing it claimed to exist for.[7]

A restorative justice process works like this. The victim and the offender agree, voluntarily, to meet. This is the essential condition, and it cannot be overstated. No one is forced into the room. Both parties must choose to be there, knowing what the process will ask of them. A trained facilitator guides the conversation. The victim describes what happened to them, and the description goes beyond the legal facts of the case into the human reality. They explain what the crime felt like, what it did to their sense of safety, how it affected their family, their sleep and their ability to walk down the street without fear.

The offender listens, and the listening happens without the buffer of lawyers or the formality of a courtroom. They sit across from the person they harmed and hear what their actions produced.[7]

Then, together, they discuss what repair might look like. Sometimes this involves financial restitution. Sometimes it involves community service directed at the type of harm the crime caused. Sometimes it involves an ongoing commitment to a specific form of accountability that the victim chooses. Almost always it involves an apology, but not the hollow performance of a courtroom statement read by a defense attorney. This is an apology shaped by the offender's genuine understanding of what they did, delivered to the face of the person who lived through it.

Lawrence Sherman and Heather Strang conducted the most comprehensive analysis of restorative justice outcomes ever assembled. Their 2007 meta-analysis examined programs across multiple countries and crime categories.[7] The findings were consistent. Restorative justice reduced reoffending across nearly every study they examined. Victim satisfaction was consistently higher than in conventional prosecution. Victims reported feeling heard, respected, and less fearful. Offenders showed greater accountability and greater understanding of the consequences of their actions.

The most counterintuitive finding was also the most important. Restorative justice was most effective for violent crimes.[7] Property offenses showed benefits, but the strongest results appeared in cases involving assault, robbery, and other crimes where the harm was deeply personal. The explanation made intuitive sense. A property crime is abstract. A violent crime is intimate. The encounter between victim and offender carries the most transformative power when the harm is most profound, because that is when the offender's ability to deny the human reality of their actions breaks down most completely.

The connection to the Ring of Gyges runs through every dimension of this process. The ring's essential function is to render the wearer invisible to the consequences of their actions. It creates a wall between the person who acts and the person who suffers the action. When Gyges put on the ring, the people he harmed became abstractions. The queen was a means to power. The king was an obstacle to be removed. Their humanity was irrelevant because their suffering was invisible to him.

Restorative justice works by removing the ring between offender and victim. The offender can no longer be anonymous to the person they harmed. But the method of removing the ring matters enormously. Surveillance removes the ring through threat. Punishment removes the ring through force. Restorative justice removes the ring through encounter.[7] It makes the other person real. It restores the human connection that the crime severed. The offender is not coerced into better behavior. They are invited to see what their behavior actually did.

New Zealand offers perhaps the most ambitious example of

restorative justice embedded in a national legal system. Since 1989, the country has required restorative conferencing for all youth offenders.[7] Every young person who commits a crime in New Zealand goes through a family group conference before any court proceeding occurs. The young offender, their family, the victim, the victim's supporters, and a youth justice coordinator sit together in a room and work through what happened. The results over three decades have been striking. Reoffending rates among youth who go through the conferencing process are significantly lower than rates among those processed through conventional courts. New Zealand's youth incarceration rates are among the lowest in the developed world.

In the United States, the Longmont Community Justice Partnership in Colorado has operated a restorative justice program since 1996. Offenders who go through the program show significantly lower rates of reoffending compared to those processed through conventional channels.[7] The program handles cases ranging from theft to assault, and the vast majority of participants, both victims and offenders, report that the process was fair.

The pattern across all these programs is the same. When you make the harmed person visible to the person who caused the harm, something changes. The change does not happen in every case or with every offender. Some people sit across from their victims and feel nothing. Some people are too damaged or too dangerous for the process to reach. The advocates of restorative justice do not claim it works universally, and intellectual honesty requires acknowledging its limits. But it works often enough that the data is clear. Encounter works where enforcement alone does not. The invisible wall between offender and victim, the wall the ring creates, can be taken down one conversation at a time.

The Norwegian prison model builds better people through environment and habit. Restorative justice builds better people through encounter and accountability. But both of these interventions come after the harm has been done. They address the person who has already put on the ring and used it. The deeper question is whether it is possible to build people who resist the ring from the beginning.

Aristotle thought the answer was yes, and his reasoning was more practical than anything Socrates or Plato had offered.[12] Virtue, he argued, is not a form of knowledge. Socrates had claimed that people do wrong only out of ignorance, that if you truly understood the good you would pursue it. Aristotle rejected this. He observed that plenty of people understand perfectly well what is right and choose otherwise. Knowing that honesty is a virtue does not make you honest any more than knowing that fitness is healthy makes you fit. Virtue requires practice.

The analogy Aristotle returned to most often was musical training.[12] You do not become a lyre player by studying the theory of music. You become a piano player by playing the piano, badly at first, then less badly, then competently, and eventually with such fluency that the correct movements become second nature. Moral virtue works the same way. A person becomes just by doing just things, courageous by acting in situations that frighten them, honest by telling the truth when lying would be easier. Each repetition deepens the habit until the virtuous response is not a calculation but a reflex.

This process requires a community. Aristotle was emphatic on this point.[12] Virtue cannot develop in isolation. It requires other people to practice with, models of virtuous behavior to imitate, and a social context that rewards virtue while discouraging vice. The hermit on the mountain may avoid wrongdoing, but avoidance is not the same as virtue. The person who is honest because they have never been tempted is not yet honest in any meaningful sense. True virtue, Aristotle insisted, is forged in the friction of communal life.

If Aristotle is right, then the answer to Glaucon is not a philosophical argument. It is a practice of teaching program. Build the right kind of community. Provide the right kind of practice. Shape the habits before the temptation arrives. Do this work early, do it consistently, and the ring will find a weaker grip when it finally presents itself. Several educational models have taken this insight seriously, and the research on their outcomes provides evidence that Aristotle's intuition was sound.

The Montessori method, developed by Maria Montessori in the early twentieth century, structures education around self-directed learning, intrinsic motivation, and responsibility to the community. Children in Montessori classrooms choose their own activities within a prepared environment. No grades or stickers reward them, and no punishments follow wrong answers. The motivation to learn comes from within, cultivated by the structure of the environment itself. Angeline Lillard and Nicole Else-Quest conducted a rigorous comparison of Montessori students and conventionally educated students and published their findings in *Science* in 2006.[8] The Montessori children demonstrated greater social cognition, a stronger sense of fairness, and more sophisticated moral reasoning than their conventionally educated peers. These children were better at taking the perspective of other people and more likely to share resources voluntarily. The habits of internal motivation and communal responsibility that the Montessori environment practiced every day showed up in measurable differences in moral behavior.

Social-emotional learning programs take a more targeted approach to the same goal. The Collaborative for Academic, Social, and Emotional Learning, known as CASEL, has developed a framework that hundreds of schools across the United States and internationally have adopted. SEL (Social and Emotional Learning) programs teach students to recognize and manage their emotions, set positive goals, show empathy for others, establish positive relationships, and make responsible decisions. Joseph Durlak and his colleagues conducted a meta-analysis of 213 studies involving more than 270,000 students.[9] The results were unambiguous. Students in SEL programs demonstrated improved prosocial behavior, better academic performance, and greater emotional well-being. They were less likely to engage in conduct problems and more likely to resolve conflicts constructively. The effects persisted years after the programs ended. The habits, once formed, endured long past the classroom that created them.

Perhaps the most striking example comes from Japan, where a practice called *tokkatsu* has been embedded in the national education system for decades.[11] Tokkatsu is not a single class or program. It is a philosophy of education that permeates the school day. Japanese students clean their own classrooms and hallways, serve lunch to each other, organize their own class meetings, resolve conflicts through group discussion, and plan school events collectively. The teacher does not impose order from above. The students build it together.

Ryoko Tsuneyoshi's comparative research documented the results of this approach.[11] Japanese students show remarkably high levels of collective responsibility, self-governance, and prosocial behavior. They take ownership of shared spaces and resolve disagreements through discussion rather than authority. A visitor to a Japanese elementary school will see six-year-olds mopping floors, organizing shoe racks, and distributing rice to their classmates with the practiced efficiency of someone who has done it hundreds of times. No adult stands over them. No reward is offered. The activity is simply part of what school means. The habits of communal responsibility that tokkatsu cultivates produce young people who are practiced in the daily work of cooperation and mutual obligation.

The common thread running through all of these educational models is the same thread that runs through Halden Prison and restorative justice programs. Every system that produces trustworthy people emphasizes internal motivation over external enforcement. Montessori does not bribe children into learning. SEL programs do not scare students into empathy. Tokkatsu does not force Japanese children to clean their schools at the point of punishment. All three systems create environments where the desired behavior is practiced until it becomes habitual. The person, not the cage, is the project.

This is Aristotle's insight made operational. You do not build a person who resists the ring by constructing a better argument against using it. You build that person by giving them years of practice in environments where the ring is available but unnecessary. The child who learns to share resources without being coerced is practicing for the adult who will return the wallet without being watched. The student who cleans a hallway that no teacher is inspecting is rehearsing for a life in which many of the most important moral choices occur in private.

In the second chapter of this book, five philosophers took their turns answering Glaucon's challenge. Socrates claimed that the just soul is healthier and happier, regardless of external reward. Aristotle insisted that virtue is a habit built through practice in community. Hobbes countered that humans cannot be trusted and the state must take the ring away by force. Kant maintained that morality is unconditional and the ring is irrelevant to duty. Nietzsche dismissed them all by declaring that the ring reveals the fiction of morality itself.

Centuries of argument produced no consensus. Each thinker offered a compelling framework, and each framework left something essential unexplained. The question of what people would do with the power of invisibility remained unresolved because it was, at bottom, an empirical question masquerading as a philosophical one.

The preceding chapters of this book have provided some of the empirical evidence. People with the ring do terrible things. They troll, harass and deceive. Corporations with the ring exploit, pollute and defraud. Governments with the ring surveil, torture and lie. Glaucon's intuition was not wrong. Invisibility enables cruelty on a scale he could not have imagined.

But the previous chapter showed that Glaucon's prediction was also incomplete. Most people return the wallet, pay for the bagel, and share the money even when they do not have to. There is something inside the human animal that resists the ring, something that persists even in the dark.

This chapter has tried to identify what strengthens that resistance, and the evidence points in a consistent direction. High-trust societies achieve trust not through surveillance but through institutions that reduce desperation and inequality.[1],[2],[3] When people have enough, the ring's promise of more loses some of its power. When institutions function fairly, the ring's promise of circumventing them loses some of its appeal.

Norwegian prisons reform their inmates not through punishment but through practice and community.[4],[5] Inmates who are treated as capable of change and given the tools to practice that change return to society as different people. The ring is available to every released prisoner. Eighty percent of them leave it alone.

Restorative justice repairs the damage of crime not through enforcement but through encounter.[7] When the person who caused the harm sits across from the person who suffered it, the ring of anonymity dissolves. The offender cannot un-see the human being they injured. That encounter changes behavior more reliably than any sentence imposed by a judge.

Education builds moral people not through lectures about ethics but through years of practiced experience.[8],[9],[11] Children who learn to cooperate, share, and take responsibility in environments that trust them grow into adults whose moral reflexes are deeply grooved. The child who cleaned a hallway in Tokyo becomes the adult who returns a lost wallet in Osaka. The pattern is unmistakable. Aristotle idea lean towards being right.[12]

He was closer to the truth than anyone who came before or after him. Virtue is not knowledge. It is not fear. It is not obedience to a universal law that operates regardless of context. Virtue is a habit, developed through repetition, in community, over time.

The answer to Glaucon is not a better argument. It is a better practice. But there is an honest admission that must follow this conclusion, and it is the most uncomfortable practical truth in this book.

Building virtuous people is slow. The work requires years of sustained investment in education, beginning before children can articulate what virtue means. It demands generations of institutional commitment to fairness and transparency, along with political will to maintain social safety nets that reduce desperation. Above all, it requires patience with a process whose results are not visible for decades, whose benefits cannot be demonstrated in a single election cycle.

The ring is fast. Anyone can put it on right now, use it today, profit from it this quarter. A troll can destroy someone's reputation in an afternoon. A politician can exploit fear of the other in a single news cycle. A corporation can cut corners in a single fiscal year. A government can classify its abuses before anyone sees them. The ring delivers its rewards immediately and exacts its costs slowly, if at all. Speed is the ring's greatest advantage, and impatience is the ring's most reliable ally.

Every system examined in this chapter requires something that modern democratic societies are increasingly unwilling to provide... sustained institutional commitment over long periods of time. Norwegian prisons work because Norway has maintained its correctional philosophy for decades while investing the resources to support it. Restorative justice works where it is implemented with fidelity over years. Montessori education works when it is practiced consistently from early childhood onward. Tokkatsu works because Japan has embedded it in a national educational culture for generations.

None of these systems produce results overnight. None of them fit into a quarterly earnings report or a campaign promise. All of them require a society to make a wager. The wager is that investing in people now will produce returns later, that building the moral capacity of citizens is more effective than building the walls to contain them. It is a bet that the slow, expensive, uncertain work of human development is a better strategy than the quick, satisfying work of enforcement and punishment.

This is the wager that high-trust societies have made and won. It is the wager that low-trust societies refuse to make, and the results are visible in every metric that measures human flourishing.

Socrates may have been right that the just soul is happier.[10] The evidence from psychology and neuroscience is at least consistent with his claim. People who behave with integrity report higher levels of well-being. People who maintain their moral commitments even when no one is watching show patterns of brain activation associated with internal reward. The just life may indeed be the better life, as Socrates insisted to Glaucon centuries ago.

But Socrates underestimated what it takes to build one. He believed that knowledge was sufficient, that understanding the good would produce good behavior. Aristotle knew better. The just soul is not discovered through contemplation. It is constructed through practice, over years, in communities that value the work. It is built by institutions that trust people before they have earned it, by schools that let children practice responsibility before they can fully understand it, by prisons that treat the worst offenders as people capable of becoming something other than what they are.

The ring of Gyges will always exist. There will always be moments of invisibility, opportunities to act without consequence, temptations available to anyone willing to reach for them. The question has never been whether the ring can be destroyed. It cannot. The question is whether we can build people who do not need it taken away from them.

The evidence says we can. The evidence also says it is the hardest thing a society can choose to do. It requires patience, money, institutional stability, and a faith in human potential that the daily news does everything in its power to undermine. It requires a willingness to invest in people who have not yet proven they deserve it, to extend trust before it has been earned, to accept that some of those investments will fail.

The alternative is a world of ever-expanding surveillance and ever-harsher punishment, a world that accepts Glaucon's premise and builds its institutions around the assumption that human beings are incapable of governing themselves. That world is easier to build. It is also one that no one should want to live in. The ring is fast. Virtue is slow. That is the challenge. That has always been the challenge.

References

[1] World Values Survey. Wave 7 (2017-2022) data on interpersonal trust.

[2] OECD. "Trust in Government" and "Society at a Glance" reports, annual editions.

[3] Wilkinson, Richard, and Kate Pickett. *The Spirit Level: Why Greater Equality Makes Societies Stronger*. Bloomsbury Press, 2009.

[4] Pratt, John, and Anna Eriksson. *Contrasts in Punishment: An Explanation of Anglophone Excess and Nordic Exceptionalism*. Routledge, 2013.

[5] Norwegian Correctional Service. Annual statistics and reports.

[6] Bureau of Justice Statistics. "Recidivism of Prisoners Released in 30 States." 2018.

[7] Sherman, Lawrence W., and Heather Strang. *Restorative Justice: The Evidence*. Smith Institute, 2007.

[8] Lillard, Angeline, and Nicole Else-Quest. "Evaluating Montessori Education." *Science* 313, no. 5795 (2006): 1893-1894.

[9] Durlak, Joseph A., et al. "The Impact of Enhancing Students' Social and Emotional Learning: A Meta-Analysis of School-Based Universal Interventions." *Child Development* 82, no. 1 (2011): 405-432.

[10] Plato. *Republic*. Books II, IV, IX. (Reeve translation, Hackett, 2004).

[11] Tsuneyoshi, Ryoko. *The Japanese Model of Schooling: Comparisons with the United States*. Routledge, 2001.

[12] Aristotle. *Nicomachean Ethics*. Books I-II. (Irwin translation, Hackett, 1999).

EPILOGUE – Moving Forward

The ring is not in a cave beneath a Lydian hillside. It is in your pocket right now, in your hand, or resting on the table beside you as you read these words. Your phone is the ring. It has been the ring since the first moment you created an account with a name that was not your own, typed something you would never say to a person's face, or scrolled past suffering with the flick of a thumb.

Your phone grants you selective invisibility. Anonymous accounts let you speak without being identified. Encrypted messages vanish after they are read. Virtual private networks route your traffic through servers on the other side of the world so that no one can trace your steps. Private browsing erases your history with the press of a button. Every day, billions of people turn the bezel inward and disappear into the architecture of the internet, moving unseen through spaces where their words carry weight but their identities carry none.

Your phone also grants you selective visibility. The camera in your pocket can record injustice and broadcast it to millions before the perpetrator has left the scene. Location sharing lets the people you love know where you are. Social media profiles project curated versions of your life into the feeds of people you have never met. Digital trails mark every purchase, every search, every message. You are simultaneously the shepherd in the cave and the king on the throne. You are both invisible and exposed, depending on which way you turn the bezel, depending on which moment of the day you inhabit.

An anonymous post is a turn of the ring. A comment left under a name no one can trace back to a living person is a step deeper into the cave. An ignored message is a choice made in the dark. The passive scroll through someone else's public grief, past images of flooded cities and burning forests and people begging for help, is a decision made by an invisible actor who will never be held to account for the act of looking away. These are the ordinary cruelties of the ring. They do not require malice. They require only the knowledge that no one is watching.

But the ring turns both ways. An anonymous donation to a stranger is also a turn of the bezel. An encrypted message sent to a friend in a country where the wrong words can produce a prison sentence is an act of invisible courage. A recording of a police officer brutalizing a citizen, captured by someone who could have looked away and chose instead to hold up the glass eye of the camera, is a use of the ring that Glaucon never considered. The ring does not dictate what you become. It reveals what you choose to become when the choosing is entirely yours.

The question was posed centuries ago, in a dialogue between a brother and a teacher, in a city that no longer exists in the form they knew. But you have something none of them had. The internet, the corporation, the surveillance state, the smartphone camera have given you data. Billions of people are wearing the ring right now, and billions of data points are flowing in from the largest uncontrolled experiment in moral psychology ever conducted. The results are arriving every day, and they are more complicated than either the optimists or the cynics predicted.

You have read what the data says, most people are not purely good or purely evil. Seventeen thousand wallets proved that Glaucon was not entirely right. People had the ring. The majority chose not to use it. But others did. The data does not show a species of angels. It shows a species of complicated actors whose behavior shifts with context and circumstance and the particular weight of the moment.

You know that context shapes behavior more powerfully than character does. The same person who donates anonymously to a food bank on Tuesday can post something vicious from a nameless account on Wednesday. An executive who coaches a youth soccer team on Saturday can sign off on an environmental violation on Monday because the corporate structure has made them invisible to the people downstream. Situations create the conditions for choices, and the ring is nothing if not a situation.

You have read what happens when communities invest in trust. The Norwegian prison system treats inmates as future neighbors and produces a recidivism rate that is a fraction of what the American system generates. High-trust societies do not achieve their trust through surveillance. They build it through investment in education, social safety nets, and the slow Aristotelian work of forming habits that persist even when no one is watching.

You have also read what happens when systems rely on surveillance instead. China's social credit system has created a population that behaves correctly in front of cameras and then finds ways around the system the moment the lens turns elsewhere. Total visibility, as we have seen, constrains behavior without transforming it, and compliance collapses the instant the watcher blinks.

The technologies of invisibility cannot be recalled. Anonymity is woven into the infrastructure of modern life. Encryption protects dissidents and criminals with the same mathematics. The camera in your pocket can document truth or fabricate lies. Legislation will not put the ring back inside the bronze horse, and no algorithm will seal the cave shut. The technology of invisibility will only become more sophisticated, more accessible, more deeply integrated into the way human beings communicate and transact and govern.

But the capacity for choosing not to use the ring will also never disappear. The question has not been answered by any philosopher, any experiment, or any technology. It has been handed to you as a daily practice...

What Will <u>You</u> Become When No One Is Watching

Made in the USA
Monee, IL
05 November 2026

68006393R00204